TWISTER TALES

UNRAVELING TORNADO MYTHS

Tornado image captured by Kelly Lange near Coleridge, Nebraska on June 17, 2014

ISBN 978-0-9856921-8-6

Printed in the United States of America

A Poem of Thanks ...

To all of my family and friends:

There is fiction,

There is fact,

There are such things that blend them together,

And then there's your love

And encouragement

Enduring in any kind of weather.

I treasure your kindness!

SCL

The clouds prepare for battle

In the dark and brooding silence

Bruised and sullen storm clouds

Have the light of day obscured

Looming low and ominous

In twilight premature

Thunderheads are rumbling

In a distant overture ...

-Neil Peart

Table of Contents

TWISTER TALES

UNRAVELING TORNADO MYTHS

STEVE LANORE

CERTIFIED BROADCAST METEOROLOGIST

Introduction

On April 27, 2011, a historic tornado swarm that some call the "Mega-Outbreak" brought twister death and destruction to the United States on a scale not seen in almost 40 years. Over $4 billion in damage was spread across half a dozen states by over 200 tornadoes, and 316 people died that day. A mere month later, Joplin, Missouri endured the single most devastating tornado in U.S. history with $2.9 billion in damage and 158 fatalities.

Fast-forward two years, a giant tornado up to a mile wide scours its way through Moore, Oklahoma on May 20, 2013. It levels 8,000 homes and businesses with wind so powerful it catapulted a 10-ton storage tank a half-mile. That's just crazy, but that's what tornadoes do. Twenty-four persons perished with $2 billion in property losses.

Just 11 days after that, the largest documented tornado in world history forms near El Reno, Oklahoma at an incredible 2.6-miles wide. A year later, a rare twin-funnel tornado demolished half of Pilger, Nebraska on June 16, 2014. All but one person survived in the town of 400, a testament to adequate warning time and perhaps a little luck too.

So it's a supercell slam-dunk, we've had more than our share of absolutely terrible tornadoes in the past few years.

When it comes to the science of tornadoes, there is plenty of solid information around, but also confused ideas where fact, fiction, and folklore mix together and muddy the meteorological waters.

This book examines some of the more common twister "tales" - beliefs about these destructive vortexes that are false, unreliable, or misleading. My goal is to increase your understanding of tornadoes and share some interesting facts and history about them, but I also hope that the tips and guidelines provided here will help to improve your storm safety.

Keep an eye on the sky.

Steve LaNore

A Word about Tornado Damage Scales: Old and New

Ted Fujita, a pioneer in tornado research and damage assessment, wrote a paper in 1971 describing a proposed tornado rating scale. This became the "F," or Fujita scale for tornado intensity and it was used until 2006.

Over the years, Doppler radars came online, many detailed aftermath studies were conducted and projects such as VORTEX and VORTEX2 gathered more field data; it became clear a "new and improved" scale was needed. In 2007 the Enhanced Fujita or "EF" scale replaced the F-scale in the United States, and it was put into practice in Canada in 2013.

The new scale uses more precise building damage indicators, for instance a brick building can usually endure stronger winds than a wood-sided home, so just noting that "the home was flattened" does not offer enough information for an accurate wind estimate. There are 28 different types and degrees of building or other damage given to determine the newer EF tornado ratings, making it a more reliable system for wind estimates.

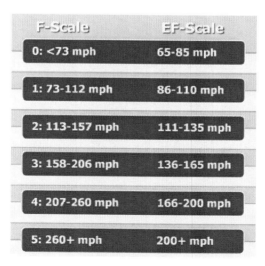

F-Scale	EF-Scale
0: <73 mph	65-85 mph
1: 73-112 mph	86-110 mph
2: 113-157 mph	111-135 mph
3: 158-206 mph	136-165 mph
4: 207-260 mph	166-200 mph
5: 260+ mph	200+ mph

Old and new Fujita tornado damage scales

Also, the classifications for tornado intensity are reduced on the Enhanced Fujita scale. Consider an F-3 tornado on the old scale where it was assigned winds of 158 to 206 mph. The new EF-scale lowers the EF-3 wind range to 136 to 165 mph, on the newer EF-scale a 200+ mph wind falls into the EF-5 category of catastrophic damage whereas a 200-mph tornado was an F-3 on the old scale. The bottom line for now is that tornadoes of EF-3 strength or higher are terrible storms that do most of the destroying and killing.

Twister Tales

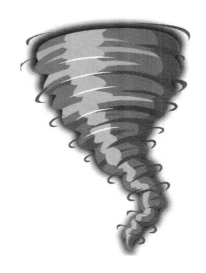

What Makes a Myth?

Some time ago I came across a story about a man who used a jar of bear grease to forecast the weather, and it turns out that he had some followers who really believed in his methods. In fact he received national exposure for his ideas back in the 1980s.

Even though this sounds kooky, there was apparently a connection between atmospheric pressure changes and the appearance of the grease … this made for occasional "success" for his forecast. I'd be happy to get a jar right now … it would save me a lot of work, but it's just not a reliable method.

On the other hand, if I properly set up a Doppler radar anywhere it will work the same way: it's a dependable and predictable tool. If something doesn't pass that test then it's not much good for the weather world. So for our purposes something doesn't have to be completely untrue to be a tornado myth, our standard will be that it's either untrue, or unreliable.

Twister Tale No. 1:
Green Skies

Myth: A green sky means that a tornado is coming.

This idea seems to show up any time there's a rash of tornadoes, with an eyewitness perhaps saying something like this: "There was a Tornado Watch in effect with storms in the area, and we happened to notice the sky turned a greenish color. So we knew a tornado was coming and took shelter, it passed down the street and just missed us."

But, is this the story everybody tells? Is a green sky truly a reliable way to predict if a thunderstorm contains a tornado? Let's take a look.

The appearance of the sky is determined by several factors:

Time of day: Sky color changes during the day, most notably around sunset and sunrise. Longer red wavelengths that are a portion of the white sunlight are scattered out in the morning and evening (adding a red tint to the sky), while a scattering of the shorter wavelengths takes place for most of the daylight hours with the result being an eye-pleasing blue background through the sky.

Cloud thickness: A 10-mile-high cumulonimbus cloud will have a much different effect on sky appearance than a small shower. Clouds filter and reduce the available light; reflection of sunlight off of nearby bright clouds can change the sky's color too.

The position of the sun relative to the cloud: If you're standing with the sun at your back and a dark cloud is in front of you, it will appear much different than if the sun is behind the storm or a cloud.

Rain and Ice: If there is rain or hail falling, it will have an influence on cloud color. However, this does not guarantee a green sky, and if the sky is green it doesn't mean that a tornado is in there. Rain and hail zones get their color from a mix of reflection, absorption, scattering, and bending of the light rays in the precipitation areas. There's a lot going on inside a storm cloud!

Particles and Moisture: The presence of dust, haze, smoke, air pollution, and even pollen are factors to consider. Smoke gives the air a murky gray or black appearance. Dust will contribute to a red, yellow, or brown color.

Very humid storm air has millions of moisture particles suspended in it that can bend and disperse the light and change its appearance to the observer. These water droplets absorb red light, making the scattered light appear blue. If this blue (scattered) light is embedded in an environment saturated in a background of red light, like at sunset, then the total effect can make the sky look greenish.

The Earth's Surface: A storm might look different over a water surface than over land, and a thunderstorm base above a white sandy beach will probably have a different hue than one over a field of bright green grass.

Consider This: Frank Gallagher is now a meteorologist for the U.S. Army, but in 1999 he conducted thesis research at the University of Oklahoma on green skies during severe weather. He used a spectrophotometer to measure the wavelength (and color) of light coming off of intense thunderstorms, and he found that the green color was real for some cases, even when the ground itself was not covered with green grass. His observations suggest that green skies are more often associated with severe weather, but not necessarily tornadoes. So this doesn't give us a lot of confidence in using green sky color as a tornado forecast tool.

Brent McRoberts of Texas A&M University offered this observation when interpreting a green sky:

"Often the sky appears almost black during a tornado, but sometimes there are greenish-looking tints to the clouds. Many tornadoes have hail right around them. What we do know for sure is that green skies do exist, but they are fairly rare. They may or may not contain a tornado, and they may or may not contain hail. We do know that they are almost always an indicator of severe weather, often very dangerous weather, so if you see a thunderstorm approaching and the sky appears green, you should take cover immediately."

So our myth microscope found that a green sky is not completely worthless as a general visual clue that a nasty storm may be coming, but it's an unreliable tornado predictor. Doppler radar and eyewitness reports are a much more dependable way to identify tornadoes or storms that might produce them than a green sky will ever be, and that makes using a green sky as a tornado forecast or warning tool a myth.

Be Weather Aware: Don't run outside to spot the storm if you are under a Tornado Warning. Take cover! If the storm is close you could be hit by flying debris, and if it's far away you probably won't be able to see it unless you live in a very open area without trees. You'll most likely just see a low cloud in the distance without being positive what it is until it might be too late.

Twister Tale No. 2:
Hot and Cold Air

Myth: Tornadoes are caused by hot and cold air meeting.

Lots of people (even some in the news media) say that "hot and cold air meeting creates tornadoes," but this is far from accurate. You can find maps similar to this one all over the Web:

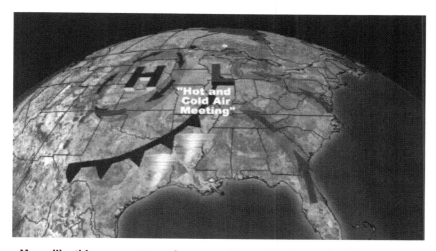

Maps like this suggest tornadoes are triggered by hot and cold air masses meeting, but this is only a small part of the truth.

But ... wind shear is what really drives the tornado machine.

Wind shear is a change of wind speed, direction, or both between two different altitudes. So, if the wind starts out at 20 mph from the south at ground level and changes to a southwest wind of 40 mph at 5,000 feet, that's considered fairly strong wind shear. It's the difference in speed between heights that tends to be most important. Since the air in this example is moving twice as fast at 5,000 feet as at ground level, the difference in forces creates a horizontal spin in the atmosphere, imagine a giant garden hose turning parallel to the ground.

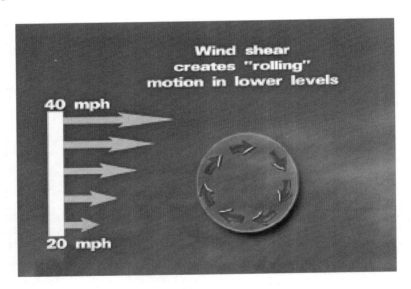

When storm updrafts start to form, this spinning horizontal air current can get stood up into the vertical. Imagine our giant garden hose pulled upward, extending tens of thousands of feet into the air – now it's a spinning updraft of air inside a developing storm. The hose has two ends, and our vertical updraft will have two parts spinning in opposite directions. The counterclockwise turning updraft usually becomes the tornado producer, and thunderstorms that produce large tornadoes develop a spin inside the cloud first before the twister forms.

A large rotating thunderstorm is called a supercell. About one supercell in four produces a tornado, and supercells account for almost all large tornadoes. The tornado is really nothing more than a violently spinning updraft in contact with the ground.

So what about hot and cold air making tornadoes? If there is cold air above warmer air, then the atmosphere wants to turn over until the heavier cold air is underneath the lighter, warmer air. This is called an unstable air mass. So in this sense, hot and cold air meeting can help get things started, but it's the wind shear that spins up the storm and plays a big factor in tornado intensity.

In fact, the Storm Prediction Center says the phrase "hot and cold air meeting causes tornadoes" is a gross oversimplification. Yet you can find many websites that make this claim, don't believe it.

The myth microscope finds fact mixed with a big dose of exaggeration so Twister Tale No. 2 flunks the truth test.

Be Weather Aware: Keep tabs on upcoming severe weather by checking the Storm Prediction Center (SPC) website every day for "convective outlooks," i.e., where severe weather might be expected. Monitor your favorite TV station's broadcasts, website, and mobile app for their information. Have a NOAA Weather Radio on hand 24/7 for updates that will continue when TV and cell service is not available. It will provide forecasts, special weather statements and hourly observations as well as severe weather watches and warnings.

Twister Tale No. 3:
Rivers, Lakes and Hills

Myth: A river, lake, or hill offers a tornado "shield."

This is a Swiss cheese claim: lots of holes. Let's take a look at some examples.

Rivers: The deadliest and longest-track tornado on record was the Tri-State Tornado on March 18, 1925. The tornado had no problem crossing the Mississippi River as it left Missouri and entered Illinois.

A large tornado forming over the Arkansas River near Little Rock / April 27, 2014 / Greg Johnson / tornadohunter.com

The Tri-State was a huge twister a mile wide, and it's possible that a smaller tornado might have been weakened by the cooler water over such a large river. However, the tornado "engine" is generally maintained by wind currents well above ground level so the storm inflow is less likely to be affected by a few minutes over a cooler water surface.

Further examples of this, at least three other large historic tornadoes crossed the Mississippi intact: the Amite, Louisiana Tornado of 1908, the St. Louis Tornado of 1896 and the Great Natchez Tornado of 1840. Another large tornado crossed the Missouri River during a March 1913 outbreak near Omaha, Nebraska; that one killed 25 people on the Iowa side after crossing.

Most rivers are quite narrow, meaning their alleged ability to protect is even less. Given the average forward speed of a tornado at 25 to 30 miles per hour, it won't be over the river long enough to be seriously affected by it. It's probably safe to say that larger tornadoes will, in general, be more immune to the potential effects of rivers, be they wide or narrow. A large river could have an effect on a small tornado, but that's a far cry from a tornado shield.

Lakes: The town of Goderich, Ontario is located on the shores of Lake Huron – it's a chilly place to swim - but that didn't stop an EF-3 tornado from forming over the lake and crashing through the town on August 21, 2011. One person died and damage was estimated at $150 million.

In 1899 a tornado began as a waterspout over a small lake in western Wisconsin before coming into New Richmond and growing into a deadly EF-5 twister (see Top 10 Deadliest Tornadoes in U.S. History for details on this killer). There are fewer accounts of tornadoes over lakes in the record because rivers present a barrier sometimes hundreds of miles long while most lakes are only a few miles long. It just doesn't happen as often.

What about hills … are they reliable "tornado protectors"? Not really. There are tons of eyewitness reports of tornadoes going down a hill, crossing a river or valley and going up the other side with no change in strength.

A study of weather in the Bighorn Mountains of Wyoming found that 18 tornadoes had formed there over a four-decade span with several of them quite large.

A colorful example takes place in May 1995 when a tornado crossed the Red River dividing Texas and Oklahoma. Eyewitness accounts tell of a fearsome sight: the tornado, over half a mile wide, turning red as it descended the Texas side of the valley, crossing the Red's muddy expanse before climbing up the Oklahoma side and continuing to dispense its wrath. The river had no effect on its strength, and it did not "skip" over the valley. It eventually destroyed the Michelin tire plant in Ardmore, Oklahoma before dissipating.

Other significant examples of hills failing the protection test include the Knoxville Tornado during the Super Outbreak of 1974, the Huntsville, Alabama tornado of November 1989, the eastern Tennessee outbreak of February 1993, and the Midwest outbreak of 2003. There are hundreds of other tornadoes that no doubt "did it" when no one was looking.

Here's what Harold Brooks, research meteorologist at NOAA's National Severe Storms Laboratory says about the belief that hills/valleys provide some mythical tornado barrier:

"Many towns in the central United States that have not been struck by a tornado for many years have a story about some topographic feature, usually a river, or a hill, that 'protects' the town from tornadoes. One of the most memorable of these legends was that Burnetts (sic) Mound protected Topeka, Kansas. On June 8, 1966, a violent tornado passed directly over Burnetts (sic) Mound, killing 16 people and causing $100 million in damage in Topeka." That equates to over a billion dollars today ... putting the Topeka tornado in the top 10 list for dollar damage in the United States. So much for Burnett's Mound.

On the flip side, Brooks offers this thought as well: "There appear to be areas [where fewer tornadoes form], such as the Ozark Mountains, but it is hard to draw definite conclusions. It is likely that topography could change the direction of the inflow into a thunderstorm, for instance, which could affect the storm's behavior."

A wind damage report shared at the 12th Americas Conference of Wind Engineers in 2013 looked at two big tornadoes from 2011 – the Tuscaloosa and Joplin events, and the study found hills did influence wind patterns in these monster twisters to some degree. Researchers at the University of Arkansas found that in both cases tornado damage tended to be less on hillsides facing away from the tornado path (there was still damage), and that both tornadoes studied tended to move toward higher ground when an elevation change was near their path. The study also showed these two particular tornadoes tended to "skip" over valleys as they moved along.

However, these factors seem to come into play intermittently at best.

Finally, the National Weather Service report on the April 27, 2011 Mega-Outbreak (see "Some Major Tornado Outbreaks" section) found that "there was a belief that being surrounded by mountains provided protection, that tornadoes did not cross rivers or only came from a certain direction."

The myth microscope shows us that these topographical features come up far short of a dependable tornado shield, so this twister tale is a myth. Believing it could be a dangerous tale to hang on to.

Be Weather Aware: A storm shelter or safe room is a worthwhile investment as these provide the only "shield" against tornadoes. Check with your local area government, there are sometimes programs available that will pay for a significant portion of a shelter's cost. These may be programs on the city, county, state, or even federal level.

Twister Tale No. 4:
Tornado Alley

Myth: Tornado Alley is where most tornadoes, or at least most "bad" tornadoes, happen.

Tornado Alley tends to be the place that is most often identified with tornadoes, but we'll prove here that it's just one of several tornado hot-spots. Search the Web for "Tornado Alley" and you'll probably find a map like this:

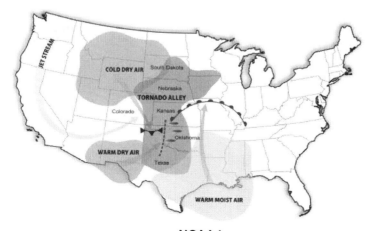

NOAA Image

Note the shaded zone extending from Texas to South Dakota. This region of the nation gets its fair share of tornadoes, some very intense and deadly, so the name is deserved.

A few examples include an EF-5 that crunched through the Oklahoma City suburb of Moore on May 20, 2013 with 8,000 structures damaged or destroyed, two dozen deaths, and $2 billion in damage. Greensburg, Kansas was leveled by a huge tornado in 2007. Moore, Oklahoma was hit by two other large tornadoes, an EF-5 in 2003 and another one in 1999. In fact, the Oklahoma City area has had 30 tornadoes of EF-3 intensity or greater since 1896!

Greensburg, KS leveled by an EF-5 on May 4, 2007 / FEMA

Andover, Kansas lost 17 lives in an EF-5 in April 1991 (11 of these were in a mobile home park). Wichita Falls, Texas endured an EF-5 in 1979. Lubbock, Texas was clobbered by a violent EF-5 tornado in 1970. And the list goes on. So in this sense you could say, "wait, these statistics prove Tornado Alley is where big tornadoes show up," and you'd certainly be right.

But ... there have been many large and catastrophic tornadoes outside of Tornado Alley, and they happen more frequently than you might think. Severe storm specialists now recognize several zones around the U.S. where tornadoes tend to form. The most notable of these "other places" is Dixie Alley.

Dixie Alley's shape on a map will depend on which website you visit, but in general it extends from Arkansas eastward through Tennessee, Mississippi, and Alabama, into Georgia. There have been many monster and very deadly tornadoes in these states through the years. The second deadliest single tornado in U.S. history hit Natchez, Mississippi in 1840 killing over 300. Of course you can argue, and rightly so, that there was no technology back then to warn the public.

Well let's fast forward 171 years. Very recently, the "Mega-Outbreak" in April 2011 produced over 200 tornadoes in one day, an all-time record. Among these was an EF-4 tornado with 190 mph winds that caused $2 billion in damage in Tuscaloosa, stayed on the ground for over 80 miles all the way to Birmingham, and was accompanied by several other giant EF-4 and EF-5 tornadoes within a 24-hour period.

It was the most concentrated tornado "swarm" since the 1930s and it happened not in Tornado Alley, but in Dixie Alley. Vilonia, Arkansas was struck twice by large tornadoes in just three years: April 25, 2011 and again on April 27, 2014. Gee, talk about bad luck! But some of it wasn't luck, it was location. Dixie Alley gets a lot of big tornadoes.

This map shows the relative density of tornadoes per year. Note that parts of Alabama, Mississippi, and Arkansas have as high a tornado incidence as central Oklahoma, Kansas, or Nebraska:

Tornado Activity in the United States
Average Number of Tornado Reports per 100 Square Miles
Reporting Years 1957-2006, F2 and Stronger

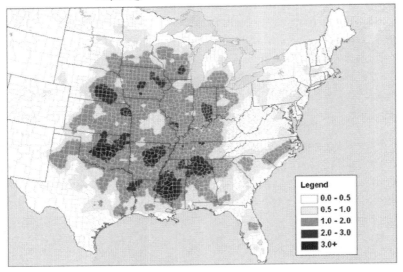

Legend
- 0.0 - 0.5
- 0.5 - 1.0
- 1.0 - 2.0
- 2.0 - 3.0
- 3.0+

SPC data

Dixie Alley storms are in some cases even more dangerous than they are in Tornado Alley. The Deep South is heavily forested in some places, making storm spotting more difficult. Dixie Alley gets its greatest number of tornadoes in early to mid-spring and also in November - when the days are shorter, making for a higher percentage of nighttime tornadoes. Overnight storms may catch residents sleeping, leaving them less chance to get to safety. More people live in mobile homes in this part of the country compared to farther west. Two cases of nighttime tornadoes tell the story: A late night outbreak (not in Dixie Alley) hit central Florida between 11 p.m. and 2:30 a.m. on February 22-23, 1998 and killed 42 people. Many of them were probably unaware that tornadoes were in the area. It was the deadliest tornado death toll in Florida history.

A larger February 2008 outbreak in Dixie Alley spawned 87 tornadoes in a 12-hour period with 57 fatalities in four states and many of these deaths were at night. One of these tornadoes, an EF-4, stayed on the ground for 122 miles through Arkansas. A noteworthy and sad statistic is that 63 percent of those killed died in mobile or manufactured homes.

Then, a massive EF-5 tornado pulverized a large swath of Joplin, Missouri on May 22, 2011. Its winds wreaked $2.9 billion in damage and claimed 158 lives. Missouri was also the starting point for the notorious Tri-State Tornado – the deadliest in U.S. history. This monster had a forward speed of over 60 mph as it killed 695 persons across parts of Missouri, Illinois, and Indiana.

This means both the deadliest and the most expensive tornadoes in the American record books took place outside of either Dixie Alley or Tornado Alley. What does this tell us?

Tornado Alley, Dixie Alley, and other tornado flash-points exist because these zones tend to be places where sufficient moisture, lifting, and wind shear come together on a regular basis to produce supercell thunderstorms: the type of storm that creates large tornadoes, but a catastrophic tornado can form anywhere conditions are right as shown later in the "Top 10 Deadliest Tornadoes in the U.S." list.

So our myth microscope debunks the Tornado Alley lock on tornadoes because there are several places around the nation where monster twisters tend to form. Tornado Alley is just the one that seems to get the most publicity.

Be Weather Aware: Nighttime tornadoes like those in the 2008 outbreak are just one example of why it's very important to have something beyond your local TV station for weather information, such as a NOAA Weather Radio or a mobile weather application on your phone for storm bulletins. What if the power goes out? Your TV won't be much good. Many television stations now stream their news broadcasts and severe weather bulletins live through their mobile apps – a great way to get information when there's no electricity, or when you are not at home. A NOAA Weather Radio with battery backup is not dependent on cell towers so it will be even more reliable for basic weather warning information than a mobile app.

Twister Tale No. 5: Tornadoes and Cars

Myth: Escaping a tornado in a vehicle is the safest move.

Myth: You should never try to flee a tornado in a vehicle.

Here we get two myths to examine for the price of one, and it's a good example of gray areas that often show up in the world of weather. These two seem to contradict each other, but that's not really the case.

It's been said for years that if a tornado is coming and you are in a vehicle, you should stop and jump into the nearest ditch or other low spot. Well this may or may not be the best idea, and in fact driving away might be a better choice in some circumstances. First, let's take a look at two examples where vehicles were tornado death traps.

The massive Wichita Falls, Texas tornado of April 10, 1979 killed 42 people, 25 of whom were in vehicles. Some were simply caught in traffic while others tried to outrun the storm. The El Reno, Oklahoma tornado of May 31, 2013 killed several veteran storm chasers who stumbled onto the EF-3 winds of a huge rain-wrapped tornado, along with two motorists who drove into it.

A study of the El Reno tornado shows that residents of the Oklahoma City area near the funnel's path had been advised by various non-Weather Service sources to get in their cars and flee from the storm. Residents were understandably on pins and needles after seeing the devastation from Moore, Oklahoma just 11 days before, but this was still not the best course of action for most people. Here's why: Had the tornado moved just a bit farther east into the Oklahoma City area, the major traffic jam taking shape there could have easily served up hundreds of people to their doom. It was a very lucky break that the tornado was a metro traffic no-show.

On the flip side, there are limited instances where driving away from a tornado may be sensible choice. In general the National Weather Service (NWS) discourages people from using their cars as a tornado getaway, however, they are sensitive to the fear people may have on days with higher tornado risk, especially when the resident lives in a mobile home. The Norman, Oklahoma NWS office posted these tips on their Facebook page on May 31, 2013 several hours before the El Reno storm took shape because there was a significant risk of tornadoes in the forecast:

"If You Do Not Feel Safe From a Tornado Where You Are...

1)...and you feel the need to drive somewhere else to find better shelter, it is critical that you do not wait too late to make that critical decision.

2) If you wait until the Tornado Warning is in effect for your location, it is probably too late to be able to drive away safely!

3) If you choose to leave in your vehicle, be sure you know where you are going before you start the car. Try to let someone know you are not at home and where you are going.

4) Do not assume that public buildings are tornado shelters. Check with your local community while the sun is still shining and before storms ever develop!

5) Be sure that you are not putting yourself in more danger by driving into another storm."

Another danger lurking for motorists is the High-Precipitation (HP) supercell. These produce rain-wrapped tornadoes like El Reno, and they can happen anywhere the air mass is very humid. A rain-wrapped tornado is a great example of why you should take cover rather than run outside to spot the funnel after a warning is issued — you can't see the thing when it's inside a shield of rain. Danger!

Some common sense must apply to exceptional situations. If you are in the path of an EF-5 twister that's 15 miles away and you live in a mobile home, you could drive out of the path of the tornado and probably be better off … but only if you know where the tornado is and where you are going. It's still risky; imagine everybody else doing the same thing and then you are trapped in a most vulnerable spot, a traffic jam.

It would probably be better in this example to drive a short distance to a sturdy permanent building and seek shelter there. Naturally, if you are out in the country then you don't have the traffic worries that urban dwellers face.

To summarize, our myth microscope has detected flaws in both statements: It's not always wrong to flee a tornado in your vehicle, but it's not generally the safest move either, so both of these statements are myths; the truth is in between.

Be Weather Aware: Ultimately you have to decide for yourself what to do, basic "storm sense" safety includes:

Know where to hide.

Be able to get there quickly.

Have multiple sources for information such as TV, a mobile app and a NOAA Weather Radio (with battery).

Keep your shelter or safe room clean and ready.

Don't panic!

Don't run outside to look at an approaching tornado.

Know exactly where you are going if you decide to drive away from the storm, know where the tornado is, which way it is moving, and at what speed ... and what roads are available to make your escape.

Twister Tale No. 6:

Bridges and Overpasses

Myth: Hiding underneath a bridge/overpass during a tornado is a great idea.

This myth took hold after a widely viewed video recorded during a 1991 tornado in Kansas (see references link). The group chose to exit their vehicle and take shelter under an overpass as the funnel approached: bad idea. They survived. The truth is, they were extremely lucky and did not take a direct hit.

The dramatic footage showed the group riding out the twister without injury under a bridge. This is an example of how doing the wrong thing worked out OK. They were just plain lucky — on the outer fringe of the vortex and not in the area of highest winds, and the tornado in question was fairly small.

A viaduct or overpass is more like a debris-filled wind tunnel than a safe haven in these cases. Wind accelerates as it travels underneath bridges, increasing the risk of injury from flying objects which is a major cause of fatalities during tornadoes. So it was a good move for the individuals to seek shelter outside of their car, and a poor choice for that shelter to be underneath the bridge.

Roger Edwards of the Storm Prediction Center in Norman, Oklahoma makes the following important points about this danger:

"Deadly flying debris can still be blasted into the spaces between bridge and grade—and impaled in any people hiding there. Even when strongly gripping the girders (if they exist), people may be blown loose, out from under the bridge and into the open......The bridge itself may fail, peeling apart and creating large flying objects, or even collapsing down onto people underneath. The structural integrity of many bridges in tornado winds is unknown—even for those which may look sturdy."

So, what should you do if caught in a vehicle with a tornado bearing down on you? Park safely off of the roadway and run into a nearby permanent structure if one is available. If not, lay flat in a drainage ditch or even on the ground if you are completely in the open. This does not apply if the ditch is flooded. Several people died in the El Reno, Oklahoma tornado from flash floods after the tornado had passed, up to 8 inches of rain in just a couple of hours made for the worst-case of a tornado followed by a flood.

Naturally, if you have time, and traffic permits, it's better to stay in your car and drive away as discussed in Myth No. 5.

The myth microscope makes it very clear that hiding under a bridge during a tornado is bad news.

<u>Be Weather Aware</u>: It's ultimately your responsibility to have access to timely and accurate weather information. If you monitor your cell phone app, NOAA Weather Radio or your local TV station broadcast/website during tornado scares, the odds of finding yourself in a bad situation where you have to hide under a bridge because there's no place else are very low. Be alert and be safe.

Twister Tale No. 7:
Tornadoes Kill the Most
People?

Myth: Tornadoes are responsible for the most weather-related deaths per year.

Twisters place third or fourth, depending on whose list you use, for the average number of weather-related deaths per year. Based on a 106-year average, extreme heat/cold events are generally the worst weather killers, followed by floods. High-end events like Hurricane Katrina in 2005 and the April 2011 Mega-Outbreak of tornadoes are thrown out of the calculation because they artificially inflate the average - kind of like the kid in class who always got the highest grade on the test!

Since many of the extreme heat/cold cases are indirect deaths, they generally do not grab headlines like tornadoes and floods. An exception was a Chicago heat spell in 1995 that killed more than 700 people — many of whom were elderly folks living in places without air conditioning or adequate ventilation. That story was sadly all over the news and rightly so. All told, an estimated 1,000 people die from extreme heat/cold exposure in the U.S. annually — a far greater toll than from tornadoes.

Coming in at second place is flooding with about 140 fatalities per year. More than half of these fatalities are preventable, such as motorists meeting their end when driving through flooded roads into fast-moving water currents. Lightning and tornadoes trade back and forth for the third place spot. Each kills 60-70 persons annually in the U.S. The year 2011 saw over 500 tornado deaths, but such a twister death toll for a single year is very rare.

<u>Be Weather Aware</u>: Lightning is often dismissed as a secondary threat while tornadoes grab all of the publicity. Three quarters of all tornadoes will have warnings issued before they touch ground or shortly afterward, while lightning strikes come without warning. If you hear thunder then the lightning is usually within 10 miles and presents a hazard. When thunder is heard, take shelter indoors. Gather more weather information from TV, radio, or Web sources so that you can determine which way the storm is headed and its intensity.

Twister Tale No. 8:
Doppler Never Misses

Myth: Doppler radar always tells you if a tornado is on the ground.

To make sense of this myth, let me offer a very crash course on the Doppler Effect. Relax, there will be no quiz.

The "Doppler Effect" was discovered by Christian Andreas Doppler in 1842. He found that when a moving object emits energy, the observed speed of the energy wave depends on both the speed of the source and the observer. This means a stationary observer will experience a noticeable "phase shift" in the emitted energy as the moving object passes. Think of a train whistle changing pitch as the locomotive moves by while you are in your car wishing it would hurry up, that's a phase shift in the sound waves.

A year after Doppler's discovery, Christoph Buys Ballot conducted an experiment to see how the effect worked for sound. Here's what he did: Ballot asked some musicians to sit in an open car attached to a locomotive and play a single note on their horns, while a second group played the same note trackside on their horns (I wonder what the engineer thought of this!). Both groups noticed the pitch of each other's notes changed as the train passed, proving the Doppler Effect worked for sound waves.

The Doppler Effect

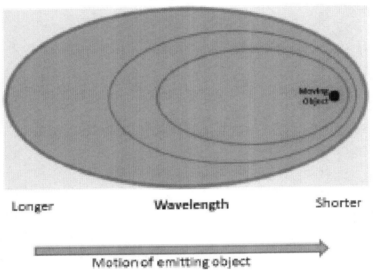

The frequency of the sound changes because when the train is approaching you have the speed of sound (760 mph) plus the speed of the train added to the sound wave's total speed, and this shortens the wavelength and the pitch is higher. The speed of the train is subtracted from the speed of sound as it moves away from you, so the wavelength is longer and the pitch drops. This is the Doppler Effect: a phase shift due to the motion of an object relative to the observer. It works for not only sounds but for all types of electromagnetic radiation including radar.

Radar uses microwave energy which travels at about 186,000 miles per second. The radar is both a transmitter and a receiver. The unit sends out very short bursts of microwaves about 1,000 times per second and then listens in between pulses for any incoming reflected energy that bounces off of rain, hail, or snow. The outgoing energy bursts are about 1.5 microseconds long, so if you crunch the numbers you will find the radar spends about 99.8 percent of its time in receive mode. The radar computer converts the return echo into a value for precipitation intensity. This is the normal reflectivity display you see where heavy rain is shown as red and light rain as green.

For the Doppler Effect, radar waves bouncing off of targets moving away from the beam will be shifted to a slightly longer wavelength, and energy coming off of particles moving toward the radar will have a slightly shorter wavelength. This is called the velocity mode and it's what allows Doppler radar to find tornadic winds inside a storm.

Velocity displays deal in two color spectrums: red and green. When several shades of red and several of green are adjacent to each in the velocity mode it indicates rotation within the storm ... and that could mean trouble.

Problem: Since the earth is curved and the radar beam travels in a straight line, the scan passes through the storm at increasing altitude as you move away from the radar site. The center of the beam slices through the storm around 12,000 feet at a range of 100 miles. The tornado is well below this level so what the radar will see at that range is a mid-level rotation that may be a part of a tornado circulation below, or the spin may be high in the storm with no tornado below. This is why spotters are so important for tornado confirmation. And there are a few gaps in the NWS' Doppler network around the nation, making it tougher to see the lowest levels of the storm where the tornado comes from.

Yet another limitation is radar beam width. It's a simple rule of geometry that a beam of energy will get wider as it moves farther from the transmitter (think of how a flashlight beam spreads out) so after 40 or 50 miles the tornado itself is too small compared to the beam width to be positively identified. The good news is that the circulation in a big supercell is much larger than the tornado itself, and extends to greater heights within the cloud. This means the more intense tornadoes, EF-3 and higher, almost always show a strong enough Doppler signature to indicate a tornado either forming or already on the ground.

A pronounced "hook echo" shows up on the radar's reflectivity ("regular") display when dealing with large tornado and supercell circulations, it's another sign of rapid rotation and tornado danger.

You might hear the term "debris ball" when reading about or watching tornado coverage on TV or the Web. The new dual-polarized NWS Doppler units can sometimes detect objects pulled into the air by the twister, and this offers very high confidence that a larger funnel is actually touching the ground.

Doppler radar is a great tool, so much so that some television stations operate their own Doppler units. These can add vital information to what's happening in the bottom of the storm, especially if the nearest National Weather Service Doppler facility is quite far away.

Our myth microscope shows us that Doppler radar alone cannot positively confirm tornadoes, because of radar beam height, beam width, and the fact that every storm is different. However, Doppler is a great tool that allows meteorologists to pinpoint and monitor storms that are already producing tornadoes (when confirmed with spotter reports) as well as those that might.

Twister Tale No. 9:

Tornado on the Ground?

Myth: A Tornado Warning means a tornado is confirmed (on the ground).

A Tornado Warning is issued based on two factors, which may be used together or separately:

1. A trained eyewitness such as law enforcement personnel, an amateur radio operator, or storm spotter sees one on the ground. Sometimes an untrained person will accurately report one too.

2. An area of strong rotation is detected by Doppler radar, and this justifies a warning. In this case, the tornado may or may not actually exist, but it is still a dangerous situation where a spinning storm may produce a tornado at any time. The Moore, Oklahoma tornado of 2013 went from the thinnest "rope" type to a monster twister a half-mile wide in just a few minutes.

Sometimes meteorologists simply can't tell if the rotation has reached the ground yet, but once the spin of the system reaches a point where it could happen, a warning is posted. This rotation begins several thousand feet above the ground, so winds underneath a developing circulation might only be 20 or 30 mph while they are spinning much faster above your head inside the cloud. The radar operator is looking for a trend. Is the storm getting better organized, is the circulation tightening and lowering, and so on.

Some small tornadoes never show up on Doppler, so again we have an example where an eyewitness would trump what the radar shows. Hmmm ... make that a dependable eyewitness, sometimes excitable folks are not sure what they're looking at and use the "T-word" far too soon.

All weather warnings are issued by your local National Weather Service office. Local media outlets relay this information to the public via TV, radio, websites, and mobile apps, but they do not issue their own warnings.

Our myth microscope uncovers a terrible truth, according to NWS reports the Tornado Warning false-alarm rate is close to 70 percent. Tornado Warnings are far from accurate, but please remember supercell storms can create dangerous winds and hail without a tornado present. This false-alarm statistic also underscores just how important trained spotters are and that our science has a long way to go.

Be Weather Aware: Treat all Tornado Warnings as "confirmed," even though there will be false alarms. There will be fewer false alarms with the larger supercells, and since these are most likely to cause fatalities, the fact is you are much more likely to get a "good" warning when a larger tornado is forming. The lower-intensity twisters lead to many of the false warnings because they are too small, weak, or short-lived to be detected.

If you have an interest in weather and would like to know how to tell real tornado clouds from look-alikes, consider attending a National Weather Service Skywarn spotter class or taking an online live webinar. These are usually held between January and April depending on the schedule from your local NWS office.

It will make you a more informed weather observer where you might be able to help alert others to a developing storm, and it's really interesting material too. More good news: these sessions are free.

Twister Tale No. 10:
Mobile Homes

Myth: Mobile homes are tornado magnets.

About 8 percent of U.S. housing units are mobile homes, but 44 percent of all tornado deaths occur in them. This number is sometimes higher in individual tornado events like the February 2008 Dixie Alley outbreak mentioned earlier where 63 percent of all deaths were in manufactured homes.

These numbers are depressing, aren't they? The problem is a mobile home is simply not designed to withstand high winds, and even "tied down" trailers are sometimes wrenched free, flipped over, pierced by flying debris or simply crushed. All of these turn the mobile home into a very poor storm hideout. So the reason there are so many tornado deaths in trailers isn't because they attract tornadoes, it's because the contest between a trailer and tornado is very one-sided.

<u>Be Weather Aware</u>: Never seek shelter in a mobile home during a Tornado Warning. Here's what the National Weather Service says to do instead: "Mobile homes are particularly vulnerable to over-turning during strong winds and should be evacuated when strong

winds or tornadoes are forecast. Trailer parks should have community storm shelters. If there is no shelter (a permanent structure or below-ground refuge) nearby, leave the trailer park and take cover on low-protected ground."

Know the best place to seek shelter for a variety of situations, and keep tabs on the weather when there's severe weather potential. Drive away from the mobile home if you have time, and seek shelter in a permanent structure. If there's no time, Get out! Lie outside in the lowest spot possible. Do not hide underneath the home.

Twister Tale No. 11:
Tornadoes and Windows

Myth: Open your windows if a tornado is coming.

This old myth is still around today, perhaps because on the surface it seems to make sense and might go something like this: "The pressure drops sharply in a tornado so I should open my windows to keep this from causing the walls and windows to burst outward ("explode") from the higher pressure inside the house."

There are some problems with this idea, though. First of all, your house is not airtight or even close. There are gaps around the doors, there are vents in the roof for plumbing, range hoods, bathroom fans, and so forth. There are air ducts and attic vents. The pressure will tend to equalize, at least partially, through these gaps as the storm moves in.

More importantly, engineering studies conducted in wind tunnels show that opening a window (or a door) actually increases the risk for the home to be destroyed. The wind coming in causes the force exerted on the roof and walls from inside to dramatically increase, blowing the roof off and allowing the walls to fail as well.

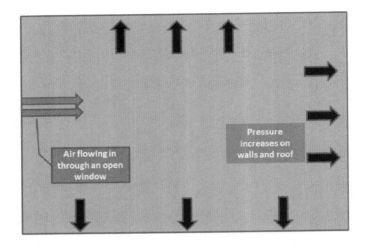

The winds in a strong tornado (EF-3 or greater) will most likely blow out the windows anyway, but for less intense storms, opening them invites flying debris, sand, mud etc. into your home. Suppose the tornado misses your house by one block: now you've survived losing your home but it's full of junk that would have stayed outside, and you have increased your chance of injury.

Then there's the risk of wasting precious time opening windows when you should be seeking safe shelter.

So our myth microscope shows that ironically, opening windows when a tornado approaches increases the chances for damage to your home!

Be Weather Aware:

1. Opening windows increases the potential for tornado winds to damage the building.

2. Opening windows invites in flying debris that would otherwise remain outside for a "near miss" or weaker tornado event, and delays your ability to take cover.

Twister Tale No. 12:
Waterspouts

Myth: Waterspouts are not dangerous.

This myth probably exists because there are two kinds of water-spouts. A tornadic waterspout is simply a tornado over water, and it forms in the same way a land-based tornado does: from the top down. Rotation begins in the cloud and low-level inflow responds to that, a funnel fills in due to condensation from the pressure drop and sometimes from the water itself.

A "fair weather" waterspout forms when warm water heats the air above it and the whirling moist air gradually becomes a visible funnel. Fair-weather waterspouts tend to form underneath the dark, flat bases of developing cumulus clouds in relatively calm weather during the early to mid-morning or late in the afternoon.

This type of waterspout is forming without benefit of a thunderstorm, so it's taking shape from the bottom up instead of the top down. By the time the funnel is visible, a fair weather waterspout is nearing its maximum size.

Fair-weather waterspouts are not extremely dangerous, although they can pose a threat to boats and swimmers as they may generate gusty winds up to 50 mph. A fair-weather waterspout is usually about half a football field in diameter and lasts just a few minutes.

Strong surface winds are the enemy of fair-weather waterspouts as they need weak winds to allow their circulation to take shape. They normally move little. If a waterspout comes ashore, the National Weather Service may issue a Tornado Warning as a few of the stronger ones can be a threat. More typically, a fair weather waterspout fades rapidly after landfall, and it rarely gets far from the beach. It can sure give you a bad hair day though, and they are no friend of boaters either.

A dramatic instance of a tornadic waterspout comes from Miami, Florida of all places, not exactly a tornado hot-spot.

In May 1997 a tornadic waterspout formed right along the beach at Miami with a sizable funnel before moving over the ocean and becoming an F-1 tornado (winds 73-112 mph, old scale). In fact, the video (see reference list for a YouTube link) shows the tornado getting stronger as it crosses a stretch of open water.

Considerable damage was done with downed trees and power lines, as well as mangled roofs reported. No one was hurt in this rather uncommon south Florida twister, but there are other cases where tornadoes forming over water are not weak as the myth goes. One happened in 1884 off of the coast of South Carolina and it severely damaged a ship!

All tornadic waterspouts are dangerous and they can grow into larger tornadoes without warning just as in the Miami storm.

Tornadic waterspouts can be expected when there is a larger system moving through and there may be severe weather expected. If there's a Tornado Watch posted for example, a waterspout is just as likely as a tornado over land.

A rare tornado along the Miami waterfront, 1997

<u>Be Weather Aware</u>: Get an update on the weather before you head out fishing, surfing, swimming, or diving. Never venture out into the sea during rough conditions or ahead of an advancing storm system. Always have a life jacket for each person on board, and a weather radio or other mobile information source.

Twister Tales from the Record Books

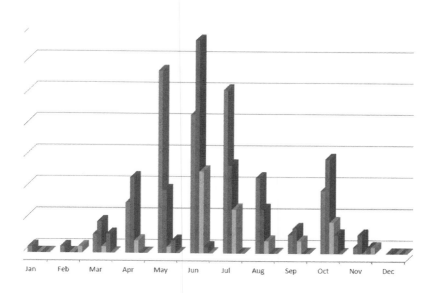

Top 10 Deadliest Tornadoes in the U.S.

It's tough to pin down what the "worst" tornado is, because numbers don't always tell the story.

A storm ranking number one on the list of deadliest tornadoes is certainly a valid reason to label it as "the worst."

However, if you are wondering which tornado left the most people homeless or caused the greatest property damage you might get different answers. Here we are avoiding a subjective term like "worst" in favor of hard numbers on how many people died as a result of these menaces.

No. 1: Tri-State Tornado

695 fatalities / March 18, 1925

The deadliest and longest-track tornado in American weather records didn't take place in Tornado Alley or Dixie Alley, but in the Midwest.

The ferocious Tri-State Tornado formed in eastern Missouri, crossed all of southern Illinois and moved into southwestern Indiana before ending its continuous 219-mile path of havoc and heartache.

The twister took shape over the hilly terrain of southeastern Missouri about 1 p.m. on March 18, 1925. This was a Wednesday so adults were at work and kids were in school.

The storm catapulted east-northeast from its initial impact point in Missouri at speeds of an incredible 60 to 70 mph (an average tornado's forward speed is 25 to 30 mph). One of the first unfortunate communities in its path was Annapolis, Missouri, where some two dozen children huddled in their brick schoolhouse while the fierce winds chewed it into rubble. On this day of great tragedy one was avoided here: by luck or by the heavens, all 25 of the children survived with only minor injuries. Hundreds of citizens downstream and in the twister's target zone would not share the Annapolis' kid's fortune.

The Tri-State Tornado clung to an unwavering path, crossing the Mississippi River and turning its fury on Gorham, Illinois, a small but thriving rail and farming town two miles from the river and in the middle of rich flood plain farmland. It laid waste to the town, killing 37, and the community was considered "100 percent destroyed," meaning no building survived without significant damage. The winds were so fierce here that railroad tracks were pulled off of the ground and flung high through the air.

Eight minutes later, Murphysboro, Illinois was in the crosshairs. This was a moderate sized city of 15,000 and was the largest population center hit by the storm. It was here that 234 persons were killed, still the highest tornado death toll for any single city in the U.S.

Fires broke out in the wake of the tornado and overall 40 percent of the city was left in ruins.

Just moments after leaving Murphysboro the fierce vortex hit De Soto, Illinois and the downtown area, where numerous citizens took shelter within a bank vault and survived, 36 who did not were killed. A 21st century example of this happened during the Moore, Oklahoma Tornado of 2013. Twenty-two people survived the Moore EF-5 mammoth by taking shelter in the Tinker Federal Credit Union's bank vault as the twister roared through.

Back to the Tri-State Tornado, as its path took it over the De Soto school where 33 children died. It's a school fatality figure that has never been exceeded by any U.S. weather event.

West Frankfort, Illinois was the next victim at 3:00 p.m. when the storm destroyed 500 buildings including many recently-built homes in this coal-mining town. Six hundred miners were safely below ground, but when they came to the surface they found their town half destroyed and 182 of their friends and neighbors dead. Some 400 were injured.

The Tri-State behemoth continued without mercy, and 10 minutes later the town of Parrish, Illinois was leveled, and just as with Gorham, Illinois, not a single structure in the town survived without damage or complete destruction. There were 36 deaths here. The huge vortex rolled on into southwestern Indiana and the town of Griffin, Indiana was all but obliterated. Residents reported seeing a multi-vortex structure, although they didn't call it that, they referred to it as "several funnels that moved around and came together." After leaving Griffin the tornado continued for a few more miles to Princeton, Indiana. In 10 more minutes the tornado would dissipate, but at this point it was still vicious enough to claim 45 lives and hurt 150 more. The deadliest tornado in American history finally dissipated over an Indiana corn field at 4:30 p.m.

The massive funnel laid waste to everything in its path. The towns of Gorham and Parrish, Illinois, and Griffin, Indiana were completely destroyed or nearly so. Annapolis, Missouri, had 90 percent of its buildings damaged or destroyed. Parrish was never rebuilt and it became a "tornado ghost town."

The tornado's "cloaked" appearance as merely a boiling black cloud and its rapid speed of movement were probably factors in the high number of fatalities, along with absolutely no warning.

Some numbers from the event include:

+ Longest-track tornado in U.S. weather records: 219 miles

+ Average width of ¾ mile, more than a mile in places, EF-5 intensity with winds of 200-plus mph

+ Three and half hours on the ground

+ Incredible forward speed of 62 mph with a maximum ground speed of 73 mph

+ 695 deaths, 2,027 injuries

+ 15,000 homes destroyed

+ $17 million in damage (1925 dollars). According to inflation conversion programs this equates to about $250 million today, although an exact comparison is very difficult because of the great changes in standards of living and technology over the past 100 years. It's probably well over a billion in real-world 21st century figures.

Imagine if the Joplin, Missouri, or Tuscaloosa, Alabama tornadoes (of 2011) had happened with no warning of any kind. That's what they faced in Murphysboro, Illinois; Griffin, Indiana, and other towns along the Tri-State Tornado's path: pockets of dense population, over 1,000 people per square mile in some spots, with no idea what was coming.

We can thank the tremendous advances in weather science, communications, and building engineering of today for keeping death tolls in even the most horrific tornadoes much lower than in the past.

No. 2: Natchez, Mississippi
317 fatalities / May 6, 1840

Called the "Great Natchez Tornado," people in its path had no chance of receiving a warning beyond a visual sighting.

The city was an important mid-19th century river port located along the east bank of the mighty Mississippi. When the tornado touched down at lunchtime it killed the majority of its victims in the port area. There were 269 deaths here before the massive mile-wide tornado slammed into town and claimed 47 more lives within the city itself.

The Natchez Free Trader newspaper offered this account of the toll on river commerce and lives: "the destruction of the flat boats is immense; at least 60 were tossed for a moment on a raging river and then sunk, drowning most of their crews. The best-informed produce dealers estimate the number of lives lost by the sinking of flat boats at 200! No calculation can be made of the amount of money and produce swallowed up by the river. The Steamboat *Hinds*, with most of her crew, went to the bottom, and the *Prairie* from St. Louis, was so much wrecked as to be unfit for use. The steamer *St. Lawrence* at the upper cotton press is a total wreck."

It's significant to note that this tornado began on the Louisiana side of the Mississippi River and crossed it without skipping a beat - never easing its fierce winds along the way. This would likely be rated an EF-5 with winds of 200 mph or greater.

No. 3 St. Louis, Missouri
255 fatalities / May 27, 1896

St. Louis was a sprawling city of some half a million people by the mid 1890s, and an important river port and industrial center. A huge mile-wide tornado crashed through the heart of St. Louis around 5 p.m. and laid waste to much of the river port and downtown area. Over 8,000 structures were damaged or destroyed and 255 people perished, half of them on the eastern (Illinois) side of the river.

The St. Louis tornado was one of the first to strike a major city where telephones and electricity were widely used. The loss of these services hampered rescue and clean-up efforts, and the electrical sparks from downed power lines started many fires.

An interesting meteorological note – this storm gave us one of the first actual measurements of the extreme pressure falls within a tornado. A home barometer recorded a pressure drop of 2.41 inches (82 mb) to a level of 26.94 inches (912 mb), as the funnel passed. This is a drop of nearly 10 percent from normal sea level pressure, and is consistent with measurements taken in other significant tornadoes since that time.

This event offers another example of a tornado crossing the mighty Mississippi in a similar fashion to the Tri-State Tornado and the Natchez Tornado. In fact, about half of the deaths occurred not in St. Louis but across the river in East St. Louis on the Illinois side. As in the Natchez storm, many boats were sunk along or in the river and even a bridge built to be "tornado proof" lost 300 feet of its span.

According to NOAA, this tornado would probably be rated an EF-4 on today's Enhanced Fujita scale with winds between 166 and 200 mph.

No. 4 Tupelo, Mississippi
216 fatalities / April 5, 1936

The hometown of Elvis Presley was devastated by a large tornado when the future star was just 16 months old. Although he survived, 216 people did not. The tornado touched down after sunset at 8:30 p.m. and moved through the western and northern parts of the city, laying waste to dozens of large mansions before turning its attention to an area known as "Gum Pond." Many of the dead were found in the small lake after the storm, a thoroughly terrible sight.

This tornado would likely be rated an EF-5 on today's scale with winds of over 200 mph.

This twister was part of an outbreak rather small in numbers (17 in two days) but great in the scope of damage and loss of life: over 450 persons killed and 3,500 injured. Damage estimates were $3 million in 1936 dollars, which equates to over $100 million today.

No. 5 Gainesville, Georgia
203 fatalities / April 6, 1936

A mere 12 hours after the Tupelo tornado finished its dirty work another mammoth "cyclone," as many called them back then, locked on Gainesville, Georgia as its target. The community was a bustling town with factories and a well-developed downtown area. Much of this was reduced to piles of debris after a tornado of estimated EF-4 intensity (166-200 mph) passed right through the downtown and factory district.

In fact, some citizens of the town tell of seeing two tornadoes converge on downtown. It's possible that this was a multi-vortex tornado, or that a dominant cyclonic tornado and a weaker anticyclonic tornado were both in action. Either way, it was a terrible day.

In one building alone, 70 people perished when the Cooper Pants plant was crushed and most of the workers inside were killed. This still stands as the largest single loss of life from a tornado in any one building in the United States. Another 20 people died when a department store collapsed. That may sound incredible in the 21st century world of reinforced concrete and steel, but most buildings in the 1930s were of a simpler brick or wood construction.

This eyewitness account of the Gainesville Tornado was printed in the Atlanta Constitution newspaper the following day:

"My home was damaged but none of us was hurt and immediately after the storm was over I went out in the rain to see about my neighbors, the Griggs. Their home was absolutely demolished. It lay in kindling wood scattered around for hundreds of yards. Mr. and

Mrs. Grigg and three of their children had been instantly killed... Their 4-year-old girl was found a few minutes later nearly a quarter of a mile from the house, or rather, from where the house had been."

The two-day Tupelo-Gainesville tornado event ranks number two in its 24-hour tornado fatality count (behind Tri-State). Even the Mega-Outbreak of 2011, for all of its horrific damage and suffering, did not equal the loss of life from these two twisters. Much of this difference can be attributed to stronger buildings, and much better warnings.

No. 6 Texas Panhandle to Woodward, Oklahoma

181 fatalities / April 9, 1947

The Sooner State's most infamous killer tornado didn't happen in Moore, Oklahoma City, or Tulsa, but in the small town of Woodward in the northwestern part of the state.

But Woodward was not the only town to be grizzled by a twister that day. The tornadic storm that would bring woe to Woodward began its assault through the town of Glazier, Texas which was all but destroyed. It continued northeast through Higgins, Texas before crossing into Oklahoma. There were 69 Texans dead before it left the state.

Although not as fast as the Tri-State Tornado's 62 mph, this twister was really hauling at forward speeds of 50 mph. It was an immense 1.8 miles wide at times, not as wide as the 2013 El Reno tornado, but still huge.

The tornado crashed into Woodward after dark at 8:42 p.m. – so there was no visual warning.

The northern and western sides of Woodward were severely mauled as 100 city blocks received extreme damage, with 107 persons killed in the town and about 1,000 injured. It's still the greatest number of lives ever lost in any Oklahoma tornado, and it is estimated to have been of EF-5 intensity.

Of note is another Oklahoma tornado that struck in the southeastern part of the state just two years earlier on April 12, 1945. The Antlers Tornado killed 69 and injured 353 and destroyed about 40 percent of the town with a half-mile wide damage path. Ironically, there might have been more media attention to this tragedy except that U.S. President Franklin Roosevelt died on the same day.

No. 7 Joplin, Missouri
158 fatalities / May 22, 2011

The Joplin tornado was the most devastating in U.S. history in total dollar damage at $2.9 billion. One-hundred fifty eight persons died with some 1,000 injured.

It was the first tornado since 1953 to kill more than 100 people; it was an immense and fierce twister up to a mile wide at times and packing winds of EF-5 intensity. The storm reached its peak strength while passing through heavily developed areas making it a true worst-case scenario.

The twister cut a six-mile long path through the near-south side of the city, narrowly missing downtown. It destroyed nearly 7,000 homes and hundreds of businesses and public buildings including Joplin High School, a Home Depot, a Wal-Mart, several elementary schools and churches, two fire stations, an electrical substation, two large cell phone towers and a nursing home. Nothing within the funnel's path survived unscathed, even the seven story concrete-and-steel St. John's Regional Medical Center was so badly beaten by the 200 mph winds that it had to be torn down.

According to Weather Service damage survey teams, an incredible 15,000 vehicles were also destroyed, with some thrown hundreds of yards and left "rolled into balls." Three-story apartment complexes had their top two floors removed, offering a compelling argument to always take cover on the lowest available floor during a tornado warning.

The extreme wind created strange happenings: The asphalt was peeled away from the Wal-Mart parking lot and hurled in big chunks through the air. One surveyor noted that a rubber hose was impaled into a tree, and a four-legged chair was embedded into an exterior wood and stucco wall. A two-by-four board was driven into a concrete curb yet remained intact. Pieces of simple cardboard were driven into stucco walls.

It was a tragic day and one that might call into question the National Weather Service's warning quality because of the high number of fatalities. However, all you have to do is look at the data and see that the Weather Service and related NOAA organizations were at the top of their game.

Here's the timeline:

+ The Storm Prediction Center (SPC) placed southwestern Missouri and adjacent areas under a "moderate risk" for severe weather in their 7:55 a.m. outlook for the day.

+ A Tornado Watch was issued by the SPC for the area at 1:30 p.m., in effect until 9:00 p.m. This watch included the language "explosive thunderstorm development" with a "strong tornado or two possible."

+ An afternoon update posted by the SPC at 3:48 p.m. mentioned the potential for "cyclic" tornadoes, which are single supercells that generate one tornado after another along their path.

+ A Tornado Warning was issued from the Springfield, Missouri National Weather Service office at 5:09 p.m. that included parts of northeastern Joplin.

+ Tornado sirens were sounded for three minutes, from 5:11 to 5:14 p.m.

+ A second Tornado Warning was issued at 5:17 p.m. for southwestern parts of Joplin, where the tornado would arrive 17 minutes later.

- The tornado began to enter the southwestern outskirts of the city at 5:34 p.m. A second tornado siren was activated at 5:38 p.m. for three minutes.

- The circulation tightened very quickly after the initial touch-down, and by 5:38 p.m. EF-4 winds were occurring along its path as it moved east at 20 mph. These increased to EF-5 intensity by the time it reached the hospital just a few minutes later.

Storm spotters provided numerous accounts of a multiple vortex structure early in the life of this tornado, but the circulation quickly became rain-wrapped which added to the danger as it chewed through the city. Those who did not hear the sirens or otherwise know of the emergency may have in some cases driven right into it because it was obscured by the precipitation shield.

A study conducted by the NWS in the aftermath of the twister revealed a troubling statistic for those who were aware of the tornado: A substantial number of people either ran outside to look for the funnel or continued to watch live TV coverage without taking cover after they had received the initial Tornado Warning. The report stated that "...there were numerous accounts of people running to shelter in their homes just as the tornado struck, despite significant advance warning of the risk..."

It is essential to take cover and to not go looking for the tornado. It may indeed find you in a most unpleasant fashion.

Note: Only a small portion of the tornado path was subjected to EF-5 winds. A 2013 study by the American Society of Civil Engineers says damage from the tornado would have been considerably less had homes in Joplin been built with "hurricane clips", which allow the roof to remain secured to the house through higher winds than a house without them. Their research also found that numerous homes were simply pushed off of their foundations or even picked up in the air because the frame of the structure wasn't bolted to the concrete below.

The report stated that flying debris from the many destroyed homes added to the damage potential of the tornado and that beefing up building code requirements would reduce such destruction in future storm events. The study concluded that many of the destroyed homes would probably have survived the tornado at least in part had they been built to more stringent standards. The team of engineers studied some destroyed 150 buildings across the tornado's path and found that 83 percent of them were subjected to maximum winds of EF-2 strength, 135 mph or less.

Since the 2011 tornado, Joplin's new-construction building codes have been changed to require hurricane clips on roofs and bolts attaching the frame to the foundation at four foot intervals. Similar building codes are used in hurricane-prone Florida.

No. 8 Amite, Louisiana – Purvis, Mississippi

143 fatalities / April 24, 1908

Like the Natchez, Mississippi tornado 68 years earlier, this tornado crossed the Mississippi River during its life. The tornado began rather early in the day, especially for a huge tornado like this one with EF-5 winds. It formed at 11:45 a.m. near Livingston, Louisiana and as is typical with many tornadoes it moved northeastward.

The tornado then scratched a two-mile wide path of destruction through the town. Many structures were laid to waste in Amite, Louisiana, and 29 people were killed. The monster tornado then crossed into Mississippi, slicing through Purvis and obliterating the community. A mere seven of 150 structures were left standing, with 55 fatalities.

As the tornado continued, four railroad crew workers were killed eight miles to the south of Hattiesburg, where they sought shelter in a boxcar. It was a fateful choice as the cars were thrown 150 feet and torn apart by the tornado. In all, some 770 people were injured along the storm's track. The real total was likely higher, perhaps significantly so, as many minor injuries were probably ignored—an omission still common in modern-day tornado disasters.

The Amite–Purvis tornado may have been two separate tornadoes created as the parent supercell "cycled," but given the sketchy accounts of the day there's no way to know for sure.

Note: Another devastating tornado from a separate supercell hit Natchez, Mississippi on the same day and killed 91 people. It ranks as the 21st deadliest in U.S. history, so just as in 2011 and 1936 we see evidence of an outbreak of multiple large tornadoes – unfortunately this happens every few years and it's just the luck of the draw whether they remain in the countryside or move into urban areas.

No. 9 New Richmond, Wisconsin

117 fatalities / June 12, 1899

New Richmond was a typical Midwestern town of the day and the anticipation of the 20th century just a few months away, with all of its promises of science and a better life for all. Electric lights and telephones were still a fairly new thing and there was even word of noisy, smoky vehicles that moved without benefit of either steam or horses!

Its 2,300 residents were enjoying an early summer day with the Gollmar Brothers circus in town; this made for many visitors to this community in far western Wisconsin, nestled on the eastern shore of the small Lake St. Croix.

This tragic tornado bears testament to the false idea that lakes protect towns from tornadoes, because the New Richmond tornado began as a waterspout over the lake. A light rain began around 5 p.m. and was followed by small hail, so obviously a thunderstorm was moving in. About 6:00 p.m. the clouds began to look threatening and according to author Mrs. A.G. Boehm, some remarked "A bad storm coming," then it came ashore and it moved right into town leaving a 1,000-foot wide gash of damage and death along the way. Here's Mr. Boehm's report (husband of the town's author) of what he saw, the language of the day is strangely eloquent, even when describing such tragedy: "It was six o'clock...the air around me calm, but dark and dense, which impressed me as strangely ominous. From the southwest an immense black cloud loomed...covering rapidly the firmament as it approached. The lower part touching the earth was funnel-shaped, and I knew but too well what that meant. The funnel, on closer observation, did not quite touch, but was surrounded by fire. Presently my attention was drawn towards the northwest, and there another dense black cloud was seen swiftly approaching from that direction. This one was expanding more rapidly and directly towards the zenith. In less time than it takes to relate, the two monsters met. Then what to me appeared a mighty struggle for supremacy began. The northwest cloud struck the edge of the funnel-shaped cloud from the southwest. The latter twisted and writhed like some monster in agony, then, rolling swiftly, lowered to the ground. Then and there took place one of the wildest, most awful scenes of nature. Though when I think of it now it appears to me very unnatural, I saw wagons, horses and cattle flying in the air like chaff… then caught as they were falling and whirled up again. But I waited to see no more, for it was rapidly coming our way, but rushed amidst, that terrible roaring and clouds of dust and debris for home, half a block distant. I reached it, blew out the lamps lighting, for supper was on the table. Hurriedly calling my wife, who was calm but anxious, we rushed down the cellar. I took my position in the southwest corner; my wife knelt in prayer. In a moment the cyclone was upon us, and I drew my wife towards me, covering her in my arms as well as I could. O, God! The awful roaring of that tornado! It seemed as though the whole universe was being torn to atoms. I hear it still as it tore the house from over our heads…"

The high fatality count was likely due to the crowds in the city as well as the densely populated area it moved through. Adding to the grisly scene, a fuel tank in the local hardware store exploded and a raging fire moved across portions of the storm-strewn debris. Some who had survived the tornado were killed by the fire while trapped, hoping for rescue.

No. 10 Flint/Beecher, Michigan

116 Fatalities / June 8, 1953

Ask anyone where tornadoes form, and I suspect very few will offer "Michigan" as their first choice. Tornadoes are much less common here than they are in Texas, Alabama, or Kansas but they can happen – and with great ferocity. The Flint/Beecher, Michigan Tornado rounds out our list of the top 10 deadliest in American weather history.

The shadows were getting long late on a humid June day when a half-mile wide EF-5 tornado touched down near Flushing, on the northwestern edge of Flint, Michigan and made its way eastward into the community of Beecher just before sunset. It moved steadily down the heavily populated corridor of Coldwater Road where it clawed out a half-mile wide path of almost total devastation. Many people were at home with their younger children already in bed when this tornado hit without warning. The funnel destroyed 340 homes and damaged many more along its 27-mile path. The high school was damaged and a local drive-in theater was utterly destroyed.

Along with the 116 fatalities, over 800 others were hurt. One resident who rode out the tornado told reporters she was digging gravel out of her scalp for days. Another told of finding tiny pieces of glass in his cheek for weeks. And then there's this eyewitness account of a close call with the EF-5 devastator: "My friend Bill Grant and I were walking home from another friend's house when we saw the tornado coming down Coldwater Road. I told him "There comes a tornado." He said we don't have them in Michigan. I said, "Maybe not, but there comes one." We lay there and watched the funnel go by. We saw three houses get picked up completely intact, lifted about 70 or 80 feet into the air. They slowly rotated around and started to tilt. When they were tilted about 45 degrees, they all exploded into little pieces…"

Other survivors of the storm told local TV and newspaper reporters "we didn't even know what a tornado was." Intense tornadoes are very rare in Michigan: this is one of only two EF-5 tornadoes ever documented in the state, the other one happening near Grand Rapids in 1956.

A separate storm the same day produced an intense tornado over Monroe County in southeastern Michigan. The storm crossed the lakeshore and continued as a waterspout for 30 miles over Lake Erie, which may be the longest-track waterspout on record.

Like some other of the tornadoes on our top 10 list, these two Michigan twisters were part of a larger outbreak that continued the next day to produce a violent tornado in another uncommon place: Worcester, Massachusetts. The Worcester storm comes in as the 22nd deadliest with 90 fatalities. I include it here because of some notable oddities.

The trouble began in Massachusetts in the mid-to-late afternoon on June 9, 1953 the day after the Flint Tornado.

Around 4:30 p.m., a supercell spawned a tornado that grew into a mile wide monster.

It crashed through Holden, Shrewsbury, and Worcester, Massachusetts as it followed a 46 mile-long path to the east-southeast, almost to the city of Boston.

The funnel was described by eyewitnesses as being a "huge cone of black smoke." In addition to the deaths, nearly 1,300 were injured, and it left 10,000 people homeless. The Worcester Tornado destroyed or seriously damaged at least 4,000 buildings, including well-built factories and part of the campus of Assumption College in Worcester.

It also hoisted a great amount of debris into the air and flung it a long way. The tornado dissipated just west of Boston.

Some of the material casualties thrown far from home include a music box, an aluminum trap door, and a 2-foot square piece of roofing. These items were found near the Blue Hill Observatory, just south of Boston, almost 20 miles downstream from where they were picked up by the twister. That's a long and mighty rough ride.

Total damage due to the Worcester tornado was estimated at $52,143,000 (1953 dollars). That would be in the $1 billion range today, ranking it in the top 10 most destructive tornadoes on record.

Who Gets the

Most Tornadoes?

The United States "wins" this contest hands down. About three-quarters of Earth's tornadoes strike America with some 1,264 observed in a typical year (20-year average). The record high year for U.S. tornadoes was 2008 with 1,897 while 2013 had about half that at a record low of 908. Quite a range.

Canada sees about 80-100 twisters per year and ranks second in the world. There are storm chasers in Australia who get to see a few "down under" as well with about 20 reported in an average year. Even the cool climate of England hosts a few dozen small tornadoes every year, and in fact the United Kingdom has the highest concentration of tornadoes in the world based on land area. Would you have guessed that?

Bangladesh is number three in the world for tornadoes per year behind the U.S. and Canada. This nation has the dubious distinction of having the world's deadliest tornado tragedy, one that easily exceeds the Tri-State Tornado of 1925.

A huge and extremely violent tornado took place in Bangladesh on

April 26, 1989. According to information documented by the Bangladesh Meteorological Department, this tornado moved east and eventually northeast from the city of Daultapur to Saturia. The path was about 10 miles long and a mile wide. All homes in a 3.5-square mile area were leveled. Thousands of trees were ripped out of the ground and blown away. According to the Bangladesh Observer, "The devastation was so complete, that barring some skeletons of trees, there were no signs of standing infrastructures."

Bangladesh is a nation with a very high population density and many dwellings of poor quality. It's estimated that 80,000 people were left homeless, 1,300 died, and 12,000 were injured in the world's deadliest twister.

Another tornado catastrophe on par with the Tri-State event for fatalities struck Bangladesh just seven years later. On May 13, 1996 a massive funnel killed 700 people, and along with the tornado softball-sized hail severely beat portions of the Tangail district in the central part of the nation. This was a family of tornadoes that moved to the south-southeast at about 30 mph for 50 miles. The aftermath was eerie and tragic: 65 bodies were found suspended in trees, 30,000 houses were destroyed, and 1,600 cattle were lost. Many people were blown long distances, one person was hurled a mile.

When a tornado similar in size and intensity to the 1989 Daultapur storm struck Joplin, Missouri in 2011, the death toll was 158 rather than 1,300. Certainly better building construction played a large role in the difference, along with timely warnings.

Texas has the highest average annual tornado count in the United States at 155 (20-year average), so on paper it appears that The Lone Star State is the tornado capital of the United States. Not so fast. Texas is about four times as large as Florida, which gets 66 tornadoes per year on average. This means the Sunshine State easily wins the contest over Texas in terms of the greatest average coverage of twisters, but the good news is that Florida tornadoes are seldom like the Texas, Oklahoma, or Dixie Alley versions. The National Climatic

Data Center data for 1991-2010 shows the tornado concentrations for each state - here are the top 10 states for tornadoes per 10,000 square miles per year. I think you'll find a few surprises on the list.

1. Florida ... 12.2

2. Kansas ... 11.7

3. Maryland ... 9.9

4. Illinois ... 9.7

5. Mississippi ... 9.2

6. Iowa ... 9.1

7. Oklahoma ... 9

8. Alabama ... 8.6

9. Louisiana ... 8.5

10. Arkansas ... 7.5

Maryland might seem out of place, but it's a small state so it doesn't take very many tornadoes to skew the average.

A separate study shows that Mississippi gets the largest, long-track tornadoes, almost twice what Oklahoma has logged. The Oklahoma variety seem to have favored densely populated areas in the past 20 years but the overall area covered by the "big ones" is centered in Mississippi.

Florida gets a lot of tornadoes because it encounters the most thunderstorms per square mile of any state, and of course you have to have a thunderstorm to get a true tornado. In the Sunshine State low-level wind shear tends to be generated by various sea breezes, but normally there's not a lot of upper-level support from strong jet stream winds, and this helps to keep most Florida tornadoes quite small.

Nature doesn't know where one state ends and another begins so for places like Oklahoma where the climate varies a lot from one side of the state to the other you need to look at more than just the state numbers.

For example, the majority of Oklahoma's major tornadoes occur in the western two-thirds of the state, so your odds of a significant tornado in Idabel (southeastern Oklahoma) are considerably lower than in Oklahoma City. Same goes for Texas: a tornado in Houston is likely to be small while one in the Panhandle may quite often be very large and intense.

It's worth noting that you can slice the numbers lots of different ways to get different results. What's probably most relevant is your chance of being in a major tornado.

The vast majority of tornado deaths are caused by EF-3 to EF-5 category twisters, and for that we find several states in a horse race: Tennessee is first followed by Arkansas, and then there's a five-way tie between Oklahoma, Kansas, Mississippi, Kentucky, and Illinois. This is again for the period 1991-2010.

Some Major Tornado Outbreaks

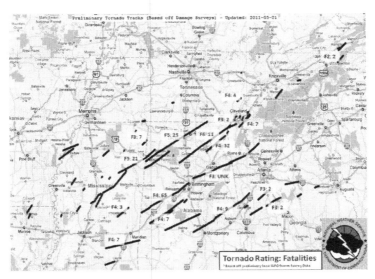

Tornado tracks on April 27, 2011 during "Mega Outbreak"
National Weather Service

Earlier we took a look at the Top 10 Deadliest Tornadoes in the U.S., in this section we look at some of the large tornado outbreaks since 1884.

A major tornado outbreak is caused by a well-defined weather system, usually one with lots of wind shear. Additionally, there's often some type of strong lifting source present such as a cold front, dry line, jet stream, upper level low, or combination of these.

Multiple significant tornadoes are the usual signature for an outbreak, and in general a bona fide "outbreak" would be composed of a minimum of 10 tornadoes, most in this list had many more than that.

They are listed from oldest to most recent. You will notice there's some overlap with the Top 10 Deadliest list, but by no means for all. This is not intended to be a list containing every outbreak, but a sampling of the larger ones and including a few with milestones in weather history. There is, in general, greater meteorological and damage detail for more recent events due to advances in technology, communication, and record-keeping.

The older outbreaks on the list have statistics that are less dependable due to poor communication and storm-damage assessment of the day, but here we will deal with what the records have to show.

The "Enigma" Outbreak

of 1884

It is sometimes referred to as the "Great Tornado Outbreak of February 1884." The name "Enigma Outbreak" was coined by Joseph Galway, a pioneer in severe storm forecasting, because of the wide range in reported deaths.

It's difficult to know the precise number of tornadoes that took place in this oldest event on our list, but the casualties and property losses paint a picture of easily more than 100 twisters.

The tornado scenario probably took shape when a significant jet stream disturbance and surface low aligned, similar to those present during the 1974 "Super Outbreak" (mentioned later).

The event began during the late morning in Mississippi and Alabama then shifted rapidly east-northeastward into Georgia and the Carolinas during the afternoon, it reached Virginia by nightfall.

At least one of the tornadic thunderstorms moved out over the waters of the Atlantic, where a waterspout seriously damaged the schooner "Three Sisters" off the coast of Charleston, South Carolina – further proof that waterspouts are dangerous, as discussed earlier.

John P. Finley was an early tornado researcher who worked for the U.S. Signal Corps. Finley stated in an 1885 report that there were a total of 44 individual tornadoes in the event. A later estimate by Finley revised that figure to 60, but it's hard to know for sure given the times.

The widespread damage and death in an era with a much lower population density would point to an even higher total than that. It's probable that many tornadoes went unreported in the more remote parts of the Appalachians, but the towns of Leeds, Alabama (east of Birmingham) and Columbus, Georgia were hit hard.

The Monthly Weather Review described Leeds as a scene of "great destruction" and of homes hit by the tornado, "not even their foundations remained."

Newspaper accounts of the day list as many as 370 deaths, including 200 in Georgia and 80 in South Carolina. These figures were based on telegraph reports sent in from towns and villages along railroad routes, so fatalities taking place elsewhere may have gone unreported.

In 1887, Finley placed the total loss of life at 800, the number of injuries at 2,500, and the number left homeless at 15,000. Finley did not share his methods for arriving at these numbers, but given his experience we would think he used a good scientific procedure for getting that total.

If his report is correct, the 800 deaths would make February 19, 1884 the single deadliest tornado day in recorded history in the United States. However, there's just no way to independently verify his conclusions and since he did not document the data, this outbreak will remain an enigma … just as the name states. In any case, it was a really bad round of tornadoes.

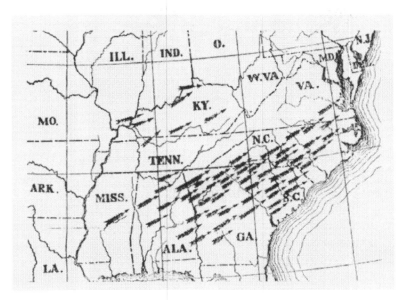

Enigma outbreak tracks as researched by John Finley.

Louisiana to Georgia

April 24-25, 1908

The Amite, Louisiana/Purvis, Mississippi tornado ranks as eighth deadliest and it was a part of this outbreak. A total of 324 died including 143 in the Amite/Purvis storm.

The long-track Amite supercell continued nearly 160 miles east-northeast into far eastern Mississippi where it finally dissipated.

Other tornado-producing storms formed over Mississippi and Alabama through the afternoon and evening. One of the Alabama storms flung a 9-ton tank a half-mile near Albertville, and all but destroyed the town of Bergens. You might recall that the Moore, Oklahoma tornado of May 2013 threw a 10-ton tank some distance, so it makes sense to estimate this particular Alabama tornado as an EF-4 or EF-5.

Galway contends that this outbreak helped spur renewed interest in accurate tornado reporting.

It seems that government bureaucracy had relegated the U.S. Weather Bureau (renamed the National Weather Service in 1970) to management under the U.S. Department of Agriculture as of 1890, and accurate recording of severe weather and tornado events suffered perhaps to some degree because of this change.

According to today's records, no organized reports of tornadoes were made between 1897 and 1907, despite a memo by the Bureau Chief Willis L. Moore following the St. Louis Tornado of 1896 that urged local offices to maintain accurate records of tornado events.

1908 marked the beginning of a 30-year period of increased tornadic activity. In 1909 there were 76 separate killer storms. A year later, 1910 became the second consecutive year with more than 400 deaths. Finally, the Bureau resumed official chronicling of tornadoes in 1916. However, truly reliable records only date back to 1950, and that year is often used as the starting point for modern tornado number-crunching.

Easter Sunday Outbreak

March 22-23, 1913

This one was rather unusual in that it took place very far to the north for March, some 300 miles farther north than the Tri-State Tornado.

Usually, northern Kansas and southern Nebraska are only slowly emerging from winter in late March and overnight lows often fall below freezing, but on March 22-23 the worst outbreak in five years rocked the nation's heartland.

The most notable tornado of the event formed late in the afternoon near Ralston, Nebraska and then moved northeastward through the western and northern sides of Omaha around 6 p.m. Seven died in Ralston.

The city of Omaha lost 94 residents to the twister in just 11 minutes along with 1,700 homes damaged or destroyed.

Another tornado moved along a parallel path from the south side of Bellevue, Nebraska (now an Omaha suburb) across the Missouri River and into Council Bluffs, Iowa. This instance is yet another offering of evidence against the myth that rivers offer protection from tornadoes - this funnel descended the river bank, crossed over and climbed the bluffs on the Iowa side before continuing into the city of Council Bluffs where it killed 25 citizens.

The same weather system spun up additional tornadoes through the night across Iowa, Missouri, and Illinois, and a few of these were also killers. To make matters even worse, very cold weather with snow came in right behind the outbreak.

A total of 191 persons perished. Many of the Iowa events were thought to be "severe windstorms," despite damage reports that pointed to tornadoes. Since most of the Iowa storms occurred after dark, the tornadoes that may have occurred were unseen.

There was erratic documentation of the Easter Sunday 1913 outbreak, and this underscores the doubt inherent in some of the nation's tornado event database before 1950.

Tri-State Tornado Outbreak

March 18, 1925

The Tri-State Tornado was documented earlier in the list for "Top 10 Deadliest Tornadoes in the U.S.," and here we discuss what might have caused such an unusual event. There were a total of eight tornadoes in the mini-outbreak of March 18, 1925, a very low number for an outbreak, but the extreme nature of the main tornado justifies its inclusion in this list. A second tornado within this outbreak killed 39 people in Kentucky and Tennessee. The immense scope of the Tri-State Tornado disaster may have led to incomplete reporting of the other tornadoes occurring on this day. The total Tri-State Tornado outbreak fatality count was 748 persons, which is the deadliest tornado day ever in the U.S.

Theories have been bandied back and forth for years on what kept the Tri-State Tornado on the ground for three and a half hours. The most likely set-up according to Storm Prediction Center forecasters John A. Hart and Robert H. Johns (retired) is one where a pocket of very high instability moved along with the supercell, allowing the tornado to continue on an extremely long track. Johns and Hart also assert that this was one tornado continuously on the ground as opposed to a "cycling" supercell with occasional tornado touch-downs or multiple supercells.

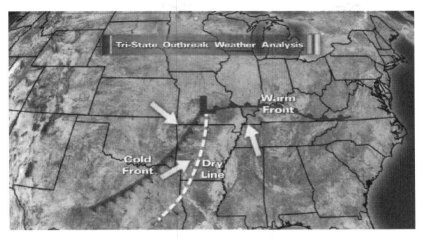

Above is the surface analysis at the approximate time of the Tri-State Tornado's formation: the twister was spawned at the intersection of a warm front, a cold front, and a dry line near a well-defined surface low.

It is unusual for the dry line to move this far to the east, and this no doubt contributed to the tornado's intensity. There was probably a significant upper trough or jet in the area too.

Deep South Outbreak

March 21-22, 1932

Ten of these tornadoes were rated "violent" (EF-5). Most of the 334 deaths occurred in Alabama, but others occurred in Georgia, Kentucky, South Carolina, and Tennessee.

The March 1932 outbreak is poorly documented in official reports. It appears that budget cuts related to the Great Depression may have adversely affected tornado tabulations. In contrast, the event was well-covered by the press.

The April 4, 1932 issue of Time magazine chronicled the tragedy wrought by the tornadoes, so some information is available.

Some of the more dramatic and horribly sad accounts include when a tornado tore an infant from its mother's arms and then dropped the baby into a well where it drowned. James Galway came across the story of a Mr. Luther Kelley, who lost his second wife to a tornado that struck Sylacauga, Alabama (near Birmingham) around sunset, leaving 1,300 homeless. Mr. Kelley's first wife had died in a tornado in the same town in 1917. How terrible! Another oddity occurred in nearby Columbiana, Alabama where a tornado killed 18 people. The storm totally destroyed a house while leaving three dozen eggs intact on a kitchen table.

Two waves of violent storms hammered Alabama. The first occurred during the afternoon, the second during the evening. The town of Northport, Alabama (near Tuscaloosa) was badly mauled with its downtown area destroyed and 100 homes flattened. Entire families were killed near Lawley, Alabama by a long-track EF-5 tornado. Twisters continued into eastern Georgia and South Carolina through the early morning hours of March 22. A total of 334 people perished.

Tennessee Valley/ N. Georgia Outbreak

April 5-6, 1936

The first week of April 1936 was a rough one for the southern United States.

A violent EF-5 tornado struck Greensboro, North Carolina on April 2, killed 13 people and damaged or destroyed nearly 300 buildings. It was the most destructive North Carolina tornado on record up to that time.

Just four days later, on April 5-6, a more intense and widespread twister outbreak spread from northern Mississippi and southern Tennessee into north Georgia and upstate South Carolina. This event remains the second deadliest outbreak on record, with at least 450 killed and approximately 2,500 injured. Most of these deaths were in two large tornadoes chronicled earlier: Tupelo, Mississippi and Gainesville, Georgia.

The trouble began brewing in Arkansas on the afternoon of April 5 when one person died in a tornado in the northeastern part of the state, but things deteriorated much more dramatically into the evening when at least three tornadoes spun up over Tennessee. One of these was an EF-5 that killed six people and cut a 35-mile path through Hardin, Wayne, and Lewis Counties. Around the same time a massive tornado cut through the western part of Tupelo, Mississippi where 216 persons died.

A temporary hospital was set up in a movie theater, and 150 railroad box cars were brought into town to shelter the homeless there.

April 6 dawned with tragedy close at hand in Georgia: a morning tornado hit the city of Gainesville. Eyewitnesses reported seeing a "pair" of tornadoes moving straight through town at 8:40 a.m. A path of catastrophic damage a quarter mile wide left hundreds of buildings destroyed and 203 dead.

This tornado was rare because intense tornadoes usually happen in the 2 p.m. to 10 p.m. time frame, every now and then overnight, but very uncommonly in the breakfast hours of the day. The 1936 Outbreak still ranks as the second deadliest behind the Tri-State Tornado Outbreak of March 1925.

Mid-South Outbreak
March 21-22, 1952

This outbreak holds a special place in weather history because it was the first event for which a Tornado Watch ... called a "Tornado Bulletin" in 1952, was issued before the tornadoes began forming. The bulletin went out around noon, so this was a really good forecast for the day, but poor communication and the violent nature of some of the tornadoes made for a day of tragedy nevertheless. There were also few if any tornado preparedness plans in those days. It's also possible the bulletin was dismissed by the public since it was something new and unproven.

Widespread tornadoes took place over parts of Arkansas, Tennessee, and adjacent states. Thirty-one tornadoes left some 175 dead and about 1,200 injured.

The first tornado touched down in the town of Dierks in southwest Arkansas around 2:30 p.m. and things picked up steam from there with several more intense twisters forming over northeast Arkansas and west Tennessee. These tornadoes continued through early the next day. The outbreak included several long-track, violent supercells, one of which devastated the north side of Henderson, Tennessee. Judsonia, Arkansas suffered nearly complete destruction, with 30 people dead and nearly 400 homes destroyed.

The good news is that the general accuracy of the forecasts for this outbreak spurred efforts to make the experimental Weather Bureau severe weather unit a permanent operation.

The outfit was renamed the Severe Local Storms (SELS) unit the following year. The Severe Local Storms unit is today known as the Storm Prediction Center (SPC), and it is the place from which se-

vere weather watches are issued, but your local NWS office issues
tornado and other severe weather warnings.

The Palm Sunday Outbreak
April 11, 1965

This outbreak was the worst ever to strike the state of Indiana. Forty-seven tornadoes caused 271 deaths and left 1,500 hurt across six states, with Indiana paying the greatest price. About half of the fatalities happened in the Hoosier State with 137 killed.

It started with a somewhat unusual layout on the weather map: As in the Tri-State outbreak 40 years earlier, a dry line was much farther east than typical, extending from a surface low over Iowa southward into Arkansas. A warm front ran eastward from the low through Illinois and Indiana, while a cold front was wrapping around the western side of the low with morning temperatures in the chilly 30s in its wake across South Dakota. An increasingly warm and humid air mass took shape through the morning ahead of these features, and a very strong jet stream would develop overhead by afternoon. The stage was set.

Fresh data arriving from weather balloons at mid-morning allowed forecasters to outline a tornado risk zone, and a public weather statement was issued at 10:45 a.m. highlighting the threat from Missouri northeastward into Indiana. The forecasters didn't have to wait long to see that it was, unfortunately, a good call. The tornadoes came in two rounds with the initial salvo hitting southern Wisconsin, northern and central Illinois, and Iowa.

Radar technology and communications had improved since the 1952 outbreak, but they had a long way to go - the first one of the day spun up at 12:45 p.m., 15 minutes before the first "Tornado Forecast" (what is today called a Tornado Watch) was issued at 1 p.m. This was a long-track EF-5 that stayed on the ground for 40 miles, mangling a number of Iowa farms in the process. One man was badly injured while trying to make it to his storm cellar. He survived the storm but succumbed to his wounds a month later. A piece of one farmhouse was hurled more than a mile by the funnel.

A tornado tracked across the city of Monroe in southern Wisconsin a little later in the afternoon where it did moderate to severe damage, but it also offered up a typical tornado oddity - it pulled the roof cleanly off of a motel while leaving the walls almost undisturbed. There was widespread damage in the extreme southern part of Wisconsin, but the only fatalities in the state took place when two vehicles were picked up and then dashed to the ground, killing all three of their occupants. A woman encountering the same tornado a bit farther along its path was very fortunate, as she was running down the basement stairs to take shelter when the house was literally transformed into a mass of splinters above her head.

Strong tornadoes were also shifting into Illinois and getting closer to Chicago. Around 3:30 p.m. a quarter-mile wide tornado touched down in the southwestern fringes of Crystal Lake (about 50 miles northwest of downtown Chicago) where extreme winds of EF-5 intensity chewed up over 150 homes in the town and killed five people. Now the massive supercell and its tornado headed east-northeastward toward Island Lake about 10 miles away.

As it approached the town it encountered a steep hill and tracked down the sharp grade to continue unhindered once reaching the bottom. The funnel remained in contact with the ground the entire time and moved through the tiny town of Island Lake, killing one and leaving more severe damage.

This is yet another case where a substantial hill offered no protection from a tornado's wrath.

The tornadic activity over Illinois ended 20 minutes later and the atmosphere went into a bit of a recess. However, a second salvo came with much greater ferocity than the first, and Indiana was squarely in the crosshairs for the worst of it. The Severe Local Storms forecasters issued another tornado watch-type bulletin this time for parts of Indiana eastward into Ohio, and including a slice of southern Michigan. Their science was solid as this is exactly where nature aimed its next barrage of tornadoes.

Numerous supercells began developing across northern Indiana as evening approached in what was a literal swarm of twisters. One particularly vicious series of tornadoes struck Elkhart, Indiana, located about 15 miles east of South Bend, or 110 miles east of Chicago.

A huge "twin vortex" tornado formed near the city about 6:30 p.m., and pictures of that extraordinary killer twister became quite famous. Here's how it looked as it moved through the countryside near Dunlap, Indiana:

Image by Paul Huffman / The Elkhart Truth / April 11, 1965

The double-funnel tornado was amazing to look at, but especially fierce as it passed through the Midway Trailer Court located in Dunlap on the southeastern edge of Elkhart. It was here that 10 persons were killed by the two vortexes. According to a paper written by famed tornado researcher Dr. Ted Fujita, this rare twin structure lasted only a minute or two and was probably the result of confused air currents within the inflow around the initial funnel. Multi-vortex tornadoes were reported in some other northern Indiana locations as dusk approached.

The Palm Sunday Outbreak produced one of the largest of its 47 tornadoes over southern Michigan, an immense two-mile wide twister around 8 p.m. Eyewitness accounts and post-storm damage reports revealed two large tornadoes moving in single file. Squarely in the path of these twisters was the Manitou Baptist Church in Lenawee County. Members of the congregation reported a small vibration turning into a low rumble that was distracting but no great cause for concern, then the sound turned into a loud roar and windows began to shatter. In a near-panic the members headed for the basement. Sadly, half of them never made it, as the tornado with winds of probably over 200 mph collapsed the building leaving some two dozen buried in the rubble. An anemometer clocked winds to 151 mph from this tornado a little farther down its path. The second tornado passed through this same misery-stricken area 40 minutes later.

Back in Dunlap, a second large storm equal in intensity to the first struck at 7:30 p.m., almost exactly one hour after the first. Whole neighborhoods were laid to waste and what was left of the little suburb was all but destroyed. A total of 60 people died in the vicinity of Elkhart.

The tornadoes continued into Ohio where it was now well after dark. More big ones were spinning up in the northwestern part of the state. One of these moved across the northern edge of Toledo with EF-5 intensity, killing over a dozen people including several in a bus that was thrown off of Interstate 75. The tornado then moved over the westernmost tip of Lake Erie where it not only assaulted cars on land but yanked boats out of the water, and sent them as missiles into the buildings along the shore.

Here's another example of the patchwork and often delayed storm information common in this outbreak: The Toledo tornado was never warned for. Ten minutes before it touched down, the Columbus weather office issued a statement saying "A moderate to heavy storm was about ten miles west of Toledo," even though the weather radars in Detroit and Fort Wayne both showed the intense parent storm on their displays, and in fact the tornado was only 15 miles from the Toledo Weather Bureau office.

The final touch-down of the outbreak was just past midnight when a tornado cut a 30-mile path through central Ohio to the southeast of Columbus. To this day, thankfully, there has not been another tornado event in this part of the country equal to the severity or loss of life experienced in the Palm Sunday Outbreak.

The fallout from this event was swift as post-outbreak investigations revealed poor communication of vital radar and storm spotter information. Additionally it was revealed that the radar network had holes in coverage and no back-up power in the event of electrical failure. Before the outbreak, several Weather Bureau offices had no access to radar except to pick up the phone and talk to Chicago, Fort Wayne or Detroit. This seems ridiculous today, but it was a different era with more primitive capabilities and much smaller Weather Service (Bureau) budgets. A concerted effort was put in place to install new radars and to make sure every Weather Bureau office had access to a live display from them.

It was also discovered that the location of teletype machines in media newsrooms and at various Weather Bureau offices was not best suited to catching weather warnings as they were issued. Also, there was no tornado siren system in most cities. It was a nice day and many people were not near a radio or TV so they had no way to learn of the warnings issued. Some Civil Defense offices did not get the warning information in a timely fashion. Another piece of fallout from the Palm Sunday Outbreak: A national Skywarn spotter training program was instituted so that tornado eyewitness reports could be quickly relayed from trained spotters to weather and emergency management officials.

Telephone communication became disrupted in at least one Weather Bureau office (South Bend, Indiana), and it was recommended that all weather bulletin communication be set up with a separate circuit over teletype to avoid this in the future.

There's no doubt these many changes reduced the loss of life in the immense "Super Outbreak" that struck farther south nine years later.

The "Super Outbreak,"

April 3-4, 1974

The "Super Outbreak" holds the No. 2 spot for the most tornadoes in any 24-hour period.

The numbers tell the story:

1. 148 tornadoes, 95 of which were EF-2 or stronger

2. 30 violent (EF-5) tornadoes and 48 killers

3. 15 tornadoes in progress simultaneously at the height of the outbreak

4. Longest-track tornado of 109 miles in Indiana

5. Storm tops measured to heights of 65,000 feet

6. 335 deaths, more than 6,000 injured

7. $600 million damage in 1974 dollars, would be several billion today, making it one of most expensive on record

8. Tornadoes occurred in 13 states and Canada.

It held the No. 1 spot for most tornadoes in a 24-hour period from 1974 until the Mega-Outbreak of 2011 exceeded this total leaving it in the No. 2 position.

The weather analysis on the morning of April 3 showed what might be considered a "classic" set-up for strong tornadoes. There was a fast-moving river of air a few thousand feet above ground level called the Low-Level Jet, and this transported a deep layer of Gulf moisture northward into the Tennessee and Ohio Valleys. A strong mid-level wave and jet streak (region of accelerated wind) was rotating over the target area from the southwest, and a surface dry line and cold front was punching eastward around the southern side of a surface low.

In many ways it was similar to the set-ups for the Tri-State and Palm Sunday Outbreaks as there was a well-defined dry line much farther to the east than usual, and this plus daytime heating, wind shear and high moisture content produced explosive thunderstorm growth. The storms began in a zone from Arkansas to Illinois, which grew and spread downstream during the mid and late afternoon.

The role of an especially moist environment and formation of the storms ahead of the cold front allowed them to remain separated from each other for a much longer period than usual, which contributed to the great number of tornadoes that took place.

Three distinct waves of tornadic storms spread grief and devastation from southern Michigan and Ontario south to the Gulf Coast states and then eastward to Virginia. The first tornado in the Super Outbreak touched down near Morris, Illinois, some 45 miles southwest of Chicago just after 2:15 p.m., and the last one finished in North Carolina at 7:00 a.m. the next morning.

Portions of metro Birmingham, Alabama; Cincinnati, Ohio; Louisville, Kentucky, and Windsor, Ontario were hard-hit.

One of the more destructive tornadoes was an EF-5 multi-vortex funnel that leveled large sections of Xenia, Ohio, killing 32 people and destroying about half of the structures in the city of 27,000. Seven of the city's 12 schools were hit but luckily school was out before the tornado arrived.

Similar massive storms struck the towns of Brandenburg, Kentucky (southwest of Louisville), and Guin, Alabama (northwest of Birmingham) with numerous deaths in both places.

A pair of violent, long-lived tornadoes tore through northern Alabama into the evening of April 3, and at times these two twisters moved over the same areas a half-hour apart, much as the Michigan tornadoes did in the Palm Sunday event.

With them came a sad tornado story of bad luck: one man, after being injured by the first storm, was killed when the church where he was taken for shelter was hit by the second tornado.

Other long-track tornadoes traveled up and down mountainsides in the southern Appalachians through the night, offering evidence that significant tornadoes are often undisturbed by rough terrain.

Multiple supercells continued to produce damaging winds and additional tornadoes until nearly dawn on April 4 as the severe weather focal point moved eastward into southern West Virginia and western Virginia. This is very rare for this time of day and also for these particular states, another testament to the off-the chart nature of this tornado outbreak.

The Super Outbreak was well-forecast. Nearly every tornado and severe weather incident that day fell within the severe weather outlook zone issued on the morning of April 3.

Most of the deaths occurred in areas under an active Tornado Watch, and most of the tornadoes were preceded by warnings. The large death toll may have been partially the fault of weaknesses in the U.S. warning system at that time and the sheer immensity of the tornadoes themselves.

For instance, timely warning communication was hampered by slow and over-burdened teletype machines and by delays that occurred within the communication relay points. There were so many warnings coming across the circuit in affected Weather Service offices that some didn't print until an hour after they were issued!

The depth of these flaws led the National Weather Service to recommend accelerated deployment of the Service's already-planned automated internal communication system. This included establishing a separate data feed for severe weather use only. A major expansion of the NOAA Weather Radio network to get warnings out quickly was also ordered. One reason: NOAA Weather Radio had proven valuable for rapid and efficient communication of storm information in places where the system was already operating, such as Atlanta, Indianapolis, and St. Louis.

University of Chicago researcher Dr. Ted Fujita and his associates conducted a thorough investigation of the Super Outbreak immediately following the event, including not only documenting tornado tracks and intensities on the ground but with aerial photography as well.

Fujita's efforts helped shed light on thunderstorm downburst winds too, because they can often occur in adjacent portions of a tornadic storm that the tornado itself misses. The damage from these microbursts might be interpreted as tornado damage to an untrained observer.

Fujita's Super Outbreak studies also further demonstrated how suction vortices, small scale circulations within a tornado, produce the narrow corridors of significantly enhanced damage often observed. These types of tornadoes are called multiple vortex tornadoes. You might think "what's the difference, it's all bad news," and this is true.

However, greater understanding of what's happening inside a severe storm leads to improved warning and forecast ability over time. That can help meteorologists more accurately model and forecast the atmosphere, aid engineers in designing sturdier buildings, and allow for better planning by emergency management officials. The research also helped develop proven techniques making for more efficient deployment of storm spotters, and ultimately saving more lives.

Great Lakes-
Atlantic States Outbreak

May 31, 1985

Dozens of violent, long-track tornadoes struck parts of Ohio, Pennsylvania, New York, and into Ontario, Canada on the afternoon and evening of Friday, May 31, 1985.

The event was the worst in recorded history over Ohio and Pennsylvania, with 10 EF-5 tornadoes and more than 30 others that claimed 88 lives and left more than 1,000 injured. Property damage in the United States alone totaled $450 million (1985 dollars).

The outbreak was spawned by a very strong jet stream and mid-level disturbance similar to the one responsible for the Flint Tornado of 1953.

Thunderstorm development was unusually explosive for this part of the country because a "cap" of warm air more typical of western Texas, Oklahoma, or Kansas was in place over the Great Lakes this day.

The warning bell sounded around 3:00 p.m. in Canada when a tornado destroyed a barn and outbuildings near Georgian Bay in southern Ontario. About an hour later, a series of tornadoes developed farther south and east. The worst of these hit the town of Barrie, killing eight people and destroying numerous homes and warehouses. Other tornadoes struck the region just north of Toronto around the same time.

In the United States, a four and a half hour binge of violent weather began around 5:00 p.m. in northeast Ohio and spread rapidly east across Pennsylvania and western New York State.

The first strike came in far northeast Ohio around 5:00 p.m., destroying mobile homes and trees. The circulation then produced EF-5 damage as it moved east into Pennsylvania and struck the town of Albion, killing nine people.

A tornado of similar intensity took shape 20 miles to the south. That storm moved east across northwestern Pennsylvania for more than 50 miles, striking the towns of Atlantic, Cooperstown, and Jamestown. Five people died in Atlantic when a factory and grain mill was destroyed.

The most destructive tornado of the outbreak took shape around 6 p.m. near Charlestown, Ohio (east of Akron). This EF-5 had a path length of nearly 50 miles and all but destroyed the town of Newton Falls as well as parts of Niles and Hubbard, where hundreds of homes were obliterated. The storm entered Pennsylvania where 95 percent of the factories in Wheatland were wiped out.

Another tornado tale: An aircraft hangar was also hit during the Wheatland storm, only for a destroyed airplane's wing to be carried 10 miles away before falling to the ground.

In a stroke of luck, a long-track supercell produced a tornado with a width of some two miles and it tracked on the ground for 70 miles – the fortunate part, it remained mainly over heavily forested Pennsylvania countryside.

Pennsylvania suffered 65 fatalities during a four-hour period on the evening of May 31, 1985. To illustrate the rarity of this event, consider that before 1985, only two tornado deaths had ever been reported in Pennsylvania in May. Eleven deaths occurred during the outbreak in Ohio, whereas only 14 tornado deaths had taken place in that state in all previous May tornadoes.

The Mega-Outbreak or Dixie Outbreak

April 27, 2011

The Mega-Outbreak exceeded the Super Outbreak of 1974 by producing an incredible 208 twisters in one day. Its peak culminated with a super-swarm of multiple mile-wide killer tornadoes.

Although technically a four-day outbreak spanning April 25-28 with an astonishing 358 tornadoes total, the bulk of the damage and fatalities in this historic event took place on April 27. Alabama bore the brunt of casualties with 62 tornadoes causing 252 deaths, thousands injured and several towns reduced largely to rubble.

It was a weather scenario forecasters recognized as "high-risk," and in fact there were three days lead time that this was going to be a major outbreak in all respects, a heads-up blessing of 21st century science not available in the 1970s during the Super Outbreak. It followed the pattern of high wind shear, strong upper jet, rich Gulf moisture tap and a surface front seen in the Palm Sunday and Super Outbreaks. But it was concentrated over a much smaller area than the Super Outbreak, making for a literal barrage of twisters through northern Mississippi and central to northern Alabama. The wind shear was downright ferocious with winds transitioning from 20 mph at the ground level to 70 mph at 3,000 feet, a rare speed at that level outside of a thunderstorm or hurricane. It created a very favorable environment for large, long-lasting twisters.

Here's what happened: A round of severe thunderstorms on the morning of April 27 swept across the Deep South. There were three deaths and a lot of straight-line wind damage, but this could not have in any way prepared southerners for what was coming in the afternoon. The morning storms also disabled several NOAA Weather Radio installations which would complicate getting warning messages out later that day.

The main batch of supercells cranked up after lunchtime over central and eastern Mississippi.

The storms triggered numerous large tornadoes, some of them long-track, as they raced east-northeast across Alabama, southern Tennessee, and Georgia through the remainder of the day and into the night.

The first tornado in nature's afternoon salvo was a vicious EF-5 that produced extreme devastation on the north side of Philadelphia, Mississippi around 2:30 p.m. Homes were scraped clean to the foundation, vehicles were thrown, and the ground itself was scoured out to a depth of two feet. The storm continued for 29 miles while more supercell thunderstorms began to rapidly form within a 150-mile wide swath of eastern Mississippi and northern Alabama.

A half-hour later, a devastating multiple-vortex tornado struck the city of Cullman, Alabama (south of Huntsville) around 3:00 p.m. The EF-4 wiped out nearly 900 homes and 100 businesses and destroyed the NOAA Weather Radio transmitter located there. Another northern Alabama tornado that began at the same time crashed through Hackleburg in Marion County. This EF-5 continued northeastward on a two-hour, 132-mile path into southern Tennessee. Hundreds of homes and businesses were destroyed, with some foundations swept completely clean. The tornado killed 72 people along its path and was at times more than one mile wide. At least 145 were reported injured.

The same tragic story was repeated again and again as this bombardment of tornadoes, the worst ever recorded (in total numbers), continued.

At 3:45 p.m., another massive tornado touched down southeast of Tupelo near the town of Smithville, Mississippi, a small community of some 900 citizens. The EF-5 wedge wiped out a police station, a post office, and half of the town's 300 homes. All but two of the downtown businesses were demolished. A pickup truck parked in front of one of the homes before the storm was never found. The NWS survey report remarked, "All appliances and plumbing fixtures in the most extreme damage path were shredded or missing."

The tornado path widened to 3/4 of a mile as it entered Alabama, where houses were destroyed near Hamilton and Shottsville. The storm caused 23 deaths and had a path length of 55 miles, making it another long-track tornado. Around this same time an EF-4 spun up about 50 miles to the south and tracked for 116 continuous miles for over two hours across the Alabama countryside and through several towns, killing 13 and injuring 54. The toll would have no doubt been much higher but warnings were nearly continuous now, with the public on edge and taking cover.

The Tuscaloosa-Birmingham tornado first appeared on the scene as it formed southwest of the city at 4:43 p.m. and proceeded to move across the city during the five o'clock news. This 190 mph EF-4 monster wrought apocalyptic ruin in its wake as it slapped away entire apartment complexes, shopping centers, and sturdy homes.

The tornado remained on the ground for an hour and a half until it reached the western and northern suburbs of Birmingham at which time the path width exceeded a mile. Nearly total devastation occurred in some areas along a track similar to that of violent tornadoes that struck the Birmingham area in 1956, 1977, and 1998. In all, the tornado remained on the ground for more than 80 miles and claimed 64 lives with another 1,500 injured. The supercell that created the Tuscaloosa tornado had a total lifespan of an incredible 380 miles and over seven hours.

The attack of tornadoes knocked out the NWS Doppler for northeastern Alabama and southeastern Tennessee located at Hytop, Alabama ... not by a direct hit but by a communications infrastructure failure. So many power lines and phone lines were down that it prevented vital radar data from getting to the NWS. The Weather Service's report on the event says this outage had a significant impact on the "warning operations" at area NWS offices and it also affected the Chattanooga, Tennessee media's ability to provide precise tornado information, because they had to use radars much farther away with less precise storm data (see Twister Tale No. 8 for more on Doppler radar).

It was a horrible day: A total of 316 lives were lost to tornadoes. Total damage was estimated at $11 billion for the complete four-day event, a U.S. record.

We can see that these outbreaks all have something in common: widespread, intense severe weather episodes covering multiple states. The systems that create these high-end outbreaks are usually spotted several days in advance now, a capability not available in the 1930s or even the 1970s.

Be Weather Aware: Monitor the Storm Prediction Center website (www.spc.noaa.gov) and/or your favorite TV meteorologist on a daily basis for a heads up on when severe weather might be in the cards. It's very unlikely with our 21st century technology that we'll miss forecasting a major tornado outbreak.

Less intense tornado events, especially for a single storm, will sometimes go un-forecast. This is why it's important that you realize a tornado can form without warning.

While less threatening than a bona fide tornado, severe thunderstorms can produce damaging wind and hail, and people are occasionally killed in severe thunderstorms. Please don't write them off.

Closing Thoughts

Tornado science and thousands of eyewitness reports over the years debunk many myths about twisters. We looked at just a few here, but the common thread is that science must be used in a way that provides predictable results or it's not going to work in the weather world. That's why green skies are not good tornado predictors and why fleeing a tornado in a car is still a gamble but may be necessary at times.

Supercell thunderstorms and the large tornadoes they produce can be spectacular and terrifying at the same time. The storm history we covered in this book tells of the incredible power, tragic outcomes, lucky breaks, and sometimes freaky things that happen when they show up.

No matter where you live, the key to surviving tornadoes or other dangerous weather is remaining weather aware - knowing the best course of action for a given location (building, mobile home, car, etc.) and choosing the safest place to hide should the forces of nature threaten your area.

And always have access to updates through multiple sources ... and know how to make the information work for you.

Keep an eye on the sky.

About the Author

Steve LaNore grew up in Florida and Texas, places that get lots of thunderstorms. This "storm exposure" began a fascination with all types of weather that led him to study meteorology at Texas A&M University where he received a Bachelor of Science degree in 1985.

LaNore has worked in several major TV markets including nine years in San Antonio and six years in Austin. He has forecasted the stormy skies of the Red River Valley of Texas and Oklahoma since 2006.

Steve holds a Certified Broadcast Meteorologist certification from the American Meteorological Society, is an advanced Skywarn spotter, and has won numerous awards for broadcast excellence.

His first book entitled "Weather Wits and Science Snickers" received three awards in 2013 including the NFPW's first place for children's non-fiction. His books "Weather Wits" and "Twister Tales" are available in paperback and e-book formats.

References

All websites were checked for function and relevance on July 15, 2014.

Introduction

Norman, Oklahoma National Weather Service
"Fast Facts for the May 20, 2013 Tornado Outbreak"
http://www.srh.noaa.gov/oun/?n=events-20130520-fastfacts

Tornado Myths

Scientific American
"Fact or Fiction? If the Sky Is Green, Run for Cover—A Tornado Is Coming" by Meredith Knight
http://www.scientificamerican.com/article.cfm?id=fact-or-fiction-if-sky-is-green-run-for-cover-tornado-is-coming

The Eagle

"Weather Whys: Hail and green skies," Department of Atmospheric Sciences, Texas A&M University
http://www.theeagle.com/brazos_life/weather_whys/weather-whys-hail-and-green-skies/article_748154de-c013-11e3-b353-001a4bcf887a.html

University of Arkansas
"New Study Shows Tornadoes Tend Toward Higher Elevations and Cause Greater Damage Moving Uphill"
http://newswire.uark.edu/articles/21786/new-study-shows-tornadoes-tend-toward-higher-elevations-and-cause-greater-damage-moving-uphill

State of Missouri
"Storm Aware-Tornado Myths"
http://stormaware.mo.gov/tornado-myths/

National Weather Service
"Tornado Myths"
http://www.crh.noaa.gov/mkx/?n=taw-part2-tornado_myths

National Weather Service
"Tri-State Tornado: A Look Back"
http://www.crh.noaa.gov/pah/1925/

Storm Prediction Center
"Significant Tornadoes in the Big Horn Mountains of Wyoming"
Jeffry Evans and Robert Johns
http://www.spc.noaa.gov/publications/evans/bighorns.htm

Stormtrack.org
Blog: "Hills and Tornadoes"
http://www.stormtrack.org/forum/archive/index.php/t-9028.html

Storm Prediction Center
"The 25 Deadliest U.S. Tornadoes"
http://www.spc.noaa.gov/faq/tornado/killers.html

Norman, Oklahoma NWS office
"Table of Tornadoes Which Have Occurred in the
Oklahoma City, Oklahoma Area since 1890"
http://www.srh.noaa.gov/oun/?n=tornadodata-okc-table

Storm Prediction Center
"Enhanced F Scale"
http://www.spc.noaa.gov/faq/tornado/ef-scale.html

National Institute of Standards and Technology (494 pages)
"Technical Investigation of the May 22, 2011 Tornado in Joplin,
Missouri"
http://nvlpubs.nist.gov/nistpubs/NCSTAR/NIST.NC-
STAR.3.pdf

The Weather Channel
"Tornado deaths prove danger of staying in cars," Associated Press
http://wxch.nl/1ddcTzG

KWTV-TV
"El Reno tornado survivor tells of others' deaths" by Nolan Clay
http://newsok.com/el-reno-tornado-survivor-tells-of-others-
deaths/article/3842601

Storm Prediction Center
"Top Ten Deadliest Tornadoes in Texas (since 1900)"
http://www.srh.noaa.gov/ama/?n=top10_tornadoes

YouTube
"Overpass Tornado"
http://www.youtube.com/watch?v=lHBZylcxIvw

University Center for Atmospheric Research (UCAR)

Atmos News, "A Terrible Tornado"

http://www2.ucar.edu/atmosnews/opinion/9696/terrible-torna-do

Storm Prediction Center

"May 31, 2013 1630 UTC Day 1 Convective Outlook"

http://www.spc.noaa.gov/products/outlook/archive/2013/day1o-tlk_20130531_1630.html

Norman, Oklahoma NWS office

"Frequently Asked Questions about the May 3, 1999 Bridge Creek/OKC Area Tornado"

http://www.srh.noaa.gov/oun/?n=events-19990503-may3faqs

The Weather Channel

"Tornado Warning False Alarms: National Weather Service Up-grades to Impact-Based Warning System" by Jon Erdman

http://www.weather.com/safety/tornadoes/tornado-warning-false-alarms-impact-based-warnings-20140418

Storm Prediction Center

"The Online Tornado FAQ"

http://www.spc.ncep.noaa.gov/faq/tornado/

Centers for Disease Control

"Heat-Related Mortality -- Chicago, July 1995"

http://www.cdc.gov/mmwr/preview/mmwrhtml/00038443.htm

NOAA Weather Partners

"U.S. Severe Weather Blog; U.S. Annual Tornado Death Tolls 1875 to present" by Harold Brooks

http://www.norman.noaa.gov/2009/03/us-annual-tornado-death-tolls-1875-present/

National Weather Service

"Thunderstorms, Tornadoes, Lightning…Nature's Most Violent Storms- A Preparedness Guide"

http://www.nws.noaa.gov/os/severeweather/resources/ttl6-10.pdf

"The Vulnerability of Mobile Home Residents in Tornado Disasters: The 2008 Super Tuesday Tornado in Macon County, Tennessee" / Philip L. Chaney and Greg S. Weaver http://journals.ametsoc.org/doi/pdf/10.1175/2010WCAS1042.1

U.S. Census Bureau

"Mobile Homes, Percent of Total Housing Units, 2007"

http://www.census.gov/statab/ranks/rank38.html

"Lessons Learned From Analyzing Tornado Damage," Timothy P. Marshall, Haag Engineering Company, Dallas, TX

http://www.agu.org/books/gm/v079/GM079p0495/GM079p0495.pdf

YouTube

"Miami Tornado of May 1997"

http://www.youtube.com/watch?v=DShp7d6yozY

NOAA Ocean Service

"A waterspout is a whirling column of air and water mist"

http://oceanservice.noaa.gov/facts/waterspout.html

National Geographic: Waterspout

http://education.nationalgeographic.com/education/encyclopedia/waterspout/?ar_a=1

Top 10 Deadliest Tornadoes
Paducah, Kentucky NWS Office

http://www.crh.noaa.gov/pah/?n=1925_tor_ss

http://www.crh.noaa.gov/pah/1925/images/trackmap.jpg

ABC News
"22 Survive Okla. Tornado by Hiding in Bank Vault"

http://abcnews.go.com/US/oklahoma-tornado-22-people-weather-storm-bank-vault/story?id=19233856

Popular Mechanics
"Tri-State Tornado: Missouri, Illinois, Indiana, March 1925" by John Galvin

http://www.popularmechanics.com/science/environment/natural-disasters/4219866

"The Great Natchez Tornado of 1840" by Stanley Nelson
http://www.natchezcitycemetery.com/custom/webpage.cfm?content=News&id=75

National Climatic Data Center
"This Month in Climate History: May 27, 1896, St. Louis Tornado"

http://www.ncdc.noaa.gov/news/month-climate-history-may-27-1896-st-louis-tornado

Digital Library of Georgia
"Tupelo-Gainesville Outbreak"

http://dlg.galileo.usg.edu/tornado/exhibit/

This Day in History: "April 5, 1936: Tornadoes Devastate Tupelo and Gainesville"

http://www.history.com/this-day-in-history/tornadoes-devastate-tupelo-and-gainesville

Digital Library of Georgia

"Effects of the Tornado on Residential Areas"

http://dlg.galileo.usg.edu/tornado/exhibit/damage/residential.php

About North Georgia

"A Time to Mourn: The Gainesville Tornado April 6, 1936" by Larry Worthy

http://www.aboutnorthgeorgia.com/ang/1936_Gainesville_Tornado

The Weather Channel

"The 10 Deadliest Tornadoes in U.S. History" by Chris Dolce

http://www.weather.com/news/tornado-central/ten-deadliest-tornadoes-united-states-20140423?pageno=5

Norman, Oklahoma NWS Office

"The Woodward Tornado of 9 April 1947"

http://www.srh.noaa.gov/oun/?n=events-19470409

National Weather Service Central Region Assessment

Joplin, Missouri Tornado (41 pages)

http://www.nws.noaa.gov/om/assessments/pdfs/Joplin_tornado.pdf

National Weather Service
Joplin Tornado Survey May 22, 2011
http://www.crh.noaa.gov/sgf/?n=event_2011may22_survey

The Joplin Globe
"Civil Engineers Release Study of Joplin Tornado Damage" by Wally Kennedy
http://www.joplinglobe.com/topstories/x120729257/Civil-engineers-release-study-of-Joplin-tornado-damage

Storm Prediction Center
"25 Deadliest U.S. Tornadoes"
http://www.spc.noaa.gov/faq/tornado/killers.html

"Noteworthy North American Tornado Outbreaks" (Power-Point slide)
Stephen Corfidi
http://bit.ly/1mL5fyD

Wisconsin Historical Society
"An eyewitness history of the New Richmond tornado, 1899"
Boehm, A.G. (c. 1900)
http://www.wisconsinhistory.org/turningpoints/search.asp?id=1543
http://content.wisconsinhistory.org/cdm/ref/collection/tp/id/65036

National Weather Service Detroit, MI
"Beecher Tornado Facts"
http://www.crh.noaa.gov/dtx/1953beecher/facts.php

Flint, Michigan Public Library

"We Remember 1953 Beecher Tornado"

http://www.fpl.info/gallery/beechertornado/index.shtml

Eyewitness report

http://www.mlive.com/news/flint/index.ssf/2008/06/beecher_tornado_memories.html

National Climatic Data Center

"U.S. Tornado Climatology"

http://www.ncdc.noaa.gov/climate-information/extreme-events/us-tornado-climatology

State of Oklahoma

"Tornadoes: A Rising Risk?" Lloyd's Insurance (31 pages)

https://www.ok.gov/oid/documents/Lloyds_TornadoRiskReport.pdf

Who Gets the Most Tornadoes?

Arizona State University

"Tornado: Deadliest Single Tornado"

http://wmo.asu.edu/tornado-deadliest-single-tornado

Storm Prediction Center

"Evidence of smaller tornado alleys across the United States based on a long track F3 to F5 tornado climatology study from 1880 to 2003"

Broyles and Crosbie

http://www.spc.noaa.gov/publications/broyles/longtrak.pdf

National Climatic Data Center
"Historical Records and Trends"
http://www.ncdc.noaa.gov/climate-information/extreme-events/
us-tornado-climatology/trends

National Climatic Data Center
"Average Number of EF0-EF5 Tornadoes per 10,000 Square
Miles, 1991-2010" (map)
http://www1.ncdc.noaa.gov/pub/data/cmb/images/tornado/
clim/avg-ef0-ef5-torn1991-2010.gif

The Washington Post
"Shock stat: Maryland has third highest tornado density in U.S."
By Jason Samenow, The Washington Post
http://www.washingtonpost.com/blogs/capital-weather-gang/
post/shock-stat-maryland-has-third-highest-tornado-density-in-
us/2012/03/26/gIQAReMmgS_blog.html

Tornado Outbreaks
National Weather Service
"Palm Sunday Tornado Outbreak Anniversary"
http://www.crh.noaa.gov/ind/?n=palmsuntor

"The Palm Sunday Story April 11, 1965"
http://www.crh.noaa.gov/iwx/program_areas/events/historical/
palmsunday1965/#Indiana%20and%20Michigan
http://www.crh.noaa.gov/ind/?n=palmsuntor

**"This Day In History: The Palm Sunday Tornadoes, April 11,
1965"** / Patrick McGuire, Elkhart County Historical Museum
Curator of Education
http://www.elkharttruth.com/living/Community-Blogs/
Elkhart-County-History/2014/04/11/This-Day-In-History-The-
Palm-Sunday-Tornadoes.html

"Palm Sunday Tornadoes of April 11, 1965", Ted Fujita and Dorothy L. Bradbury; Department of Geophysical Sciences, University of Chicago

http://docs.lib.noaa.gov/rescue/mwr/098/mwr-098-01-0029.pdf

"A Report on the Palm Sunday Tornadoes of 1965" U.S. Department of Commerce, Weather Bureau

http://www.nws.noaa.gov/om/assessments/pdfs/palmsunday65.pdf

"The Super Outbreak: Outbreak of the Century" Stephen J. Corfidi, Jason J. Levit and Steven J. Weiss NOAA/NWS/NCEP/ Storm Prediction Center, Norman, OK

http://www.spc.noaa.gov/publications/corfidi/3apr74.pdf

"Revisiting the 3-4 April 1974 Super Outbreak of Tornadoes" Corfidi, Weiss, Kain, Rabin and Levit, April 2010

http://www.spc.noaa.gov/publications/corfidi/74superoutbreak.PDF

"After 40 years, Super Outbreak remains a tornado benchmark" Stan Finger, Wichita Eagle

http://www.kansas.com/2014/03/30/3376677/after-40-years-super-outbreak.html

February 2008 "Dixie Alley" Outbreak
National Weather Service
"Service Assessment / Super Tuesday Tornado Outbreak of February 5-6, 2008" (48 pages)

http://www.nws.noaa.gov/om/assessments/pdfs/super_tuesday.pdf

National Weather Service

"Historic Tornado Outbreak April 27, 2011"

http://www.srh.noaa.gov/bmx/?n=event_04272011

National Weather Service

'Service Assessment: The Historic Tornadoes of April 2011"

http://www.nws.noaa.gov/om/assessments/pdfs/historic_torna-
does.pdf

National Severe Storms Lab

"2011 Spring Tornado Outbreaks"

https://www.nssl.noaa.gov/about/history/2011/

The Washington Post

"Super tornado outbreak of April 27, 2011: one year anniversary"
by Jason Samenow

http://www.washingtonpost.com/blogs/capital-weather-gang/
post/super-tornado-outbreak-of-april-27-2011-one-year-anniver-
sary/2012/04/27/gIQARRLJlT_blog.html

National Weather Service

"Hackleburg (Marion County) EF5 Tornado April 27, 2011"

http://www.srh.noaa.gov/bmx/?n=event_04272011hackleburg

38702721R00077

Made in the USA
Lexington, KY
21 January 2015